Getting to Know
ArcView GIS

the geographic information
system (GIS) for everyone

GeoInformation International
a division of Pearson Professional Limited
307 Cambridge Science Park
Milton Road
Cambridge
CB4 4ZD
and associated companies throughout the world

Address inquiries relating to ArcView® GIS software to Environmental Systems Research Institute, Inc., 380 New York Street, Redlands, California 92373-8100, or to your local ESRI software distributor.

First published 1996. Second edition published 1997.

British Library Cataloging-in-Publication Data
A CIP record for this book is available from the British Library.

ISBN 1 86242 019 X

Printed in the United States of America.

ArcView uses Neuron Data's Open Interface.

Contents

ENVIRONMENTAL SYSTEMS RESEARCH INSTITUTE, INC.

Part 2   Using ArcView GIS

Section 1   ArcView GIS basics

Section 4   Managing tabular data

Section 5   Analyzing spatial relationships

Preface and acknowledgments

Getting to Know ArcView GIS comes from the world leader in GIS software, Environmental Systems Research Institute, Inc. (ESRI). Founded in 1969 as a research organization to develop new methods for managing geographic information, ESRI provides software, data automation, and consulting services to thousands of GIS users around the world. ESRI's early research and development set the stage for the revolution in automated mapping that we see today.

At ESRI, our reputation is built on contributing our technical knowledge, special people, and valuable experience to the collection, analysis, and communication of geographic information. Our product, ArcView® GIS software, allows you to manage geographic information from your desktop. It's changing the way we all do business.

We would like to acknowledge the following data publishers for their respective contributions to this work:

Schools data provided courtesy of the Atlanta Regional Commission and is used herein with permission.

> The Atlanta Regional Commission (ARC) is the official planning agency for the 10-county Atlanta Region in a wide range of areas including transportation, environmental quality, land use, public facilities, job training, aging services, and other human services. It was established in 1971 to assist local governments in planning for common needs, cooperating for mutual benefit, and coordinating for sound regional development. Board membership on ARC is held by 23 local elected officials and 15 private citizens. The work of the Commission is supported by local, state, and federal funds.

City of Albuquerque data provided courtesy of the City of Albuquerque—Geographic Information System (GIS) Division and is used herein with permission.

City of Ontario data provided courtesy of the City of Ontario GIS Division and is used herein with permission.

Mata Atlantica data provided courtesy of Conservation International, Conservation International–Brazil, Conservation International–Brazil/ Universidade de Brasilia, Fundação Biodiversitas, and Sociedade Nordestina de Ecologia and is used herein with permission.

> A Conservation Priority-Setting Workshop was held in December 1993 for the Atlantic coastal forests of northeastern Brazil. The workshop utilized GIS to help determine regional conservation priorities for this species-rich area that is under severe development pressures. The data was compiled prior to the workshop and GIS was used to integrate the data, generate base maps, and analyze the synthesized information for the participating biologists and socioeconomic experts. The workshop approach provided a scientifically sound forum for determining which areas are of greatest concern for a range of conservation issues. It also generated a detailed database of the best available information on the region that will be used to support future research and decision making.

1990 and 1993 land use data provided courtesy of the Southern California Association of Governments and is used herein with permission.

Southern California Association of Governments
37th Floor
611 W. 6th Street
Los Angeles, CA 90017

Telephone: (213) 236-1801
Fax: (213) 236-1803

Redlands satellite imagery data provided courtesy of SPOT Image Corporation and is used herein with permission. Copyright © 1991, 1993 CNES. All rights reserved.

GNP per capita, population density, and birth rate data provided courtesy of The World Bank, Washington, D.C., and is used herein with permission.

Maps of downtown Portland reproduced with permission granted by Thomas Bros. Maps®. It is unlawful to copy or reproduce all or any part thereof, whether for personal use or resale, without permission. Copyright © 1995 Thomas Bros. Maps. All rights reserved.

ENVIRONMENTAL SYSTEMS RESEARCH INSTITUTE, INC.

Introduction

Move over text files, move over spreadsheets, and databases: geographic information systems have arrived on the desktop. Geographic information systems (GIS) let you visualize information in new ways that reveal relationships, patterns, and trends not visible with other popular systems.

Getting to Know ArcView GIS presents the concepts upon which this technology is based, how it works, and what it does. In part 1 of the book, you'll see how people in a wide range of fields are using desktop GIS to find potential customers, locate the best place for a new business or facility, identify natural areas needing protection, find the best places to develop real estate, manage extensive road networks, inventory forest lands, do emergency planning in urban areas, manage resources after fire and flood—the list goes on and on. You'll find out how you can use desktop GIS to study and analyze situations and create high-quality maps and charts. You also get to see, in chapter 6, an entire application using a particular desktop GIS, *ArcView GIS*. In part 2 of the book (chapters 7–29), you'll learn how to use ArcView by working through exercises that are based on real-life situations.

To supplement the material in this book, we also provide a multimedia CD–ROM. The CD–ROM has three parts. The Desktop GIS Primer mirrors the content of the book's first six chapters, presenting similar information in a dynamic way. The ArcView GIS Showcase demonstrates how ArcView software works; you'll see ArcView in action, performing the tasks described in chapters 1–6. The ArcView GIS Tutorial contains a version of ArcView that's locked to the data and exercises in the book, a help file that contains all of the exercise steps, and videos that let you see ArcView performing each of the exercise steps. We've designed the book, the Desktop GIS Primer, the ArcView GIS Showcase, and the ArcView Tutorial so that you can use them independently or together. The choice is yours.

Before you install the CD–ROM, be sure to read the license agreement in appendix C and the installation instructions in appendix D.

Throughout part 2 of the book, you'll see teal-colored boxes like this one:

> **Zooming in and out in a view.** ArcView gives you a variety of ways to zoom in and out. You can zoom from the center of the view or from a position or area you define with the mouse. For more information, search for these Help Topics: *Zooming in and out on a view, Zoom In, Zoom Out, Zoom In tool, Zoom Out tool.*

These boxes offer more detailed explanations for topics covered in the exercises. To search for a Help Topic, select "Search For Help On" from the Help menu.

Finally, if you're looking for information on how to contact ESRI, the makers of ArcView, please refer to appendix A, where you'll find the telephone and fax numbers of our offices in the United States and throughout the world.

PART 1

Getting to know desktop GIS

These first six chapters introduce you to desktop GIS and its concepts. You'll learn why desktop GIS maps are dynamic, how to get information from them, and how to use them to study relationships and analyze locations. You'll see how to use desktop GIS to create quality presentations, including maps, charts, images, and more. You'll find out where you can get data and how to evaluate it. You'll even learn how to create some data of your own. Finally, you'll see how related GIS tasks are performed, step-by-step, using a real desktop GIS.

Desktop GIS:
What it is
and what it does

The new source of power is not

money in the hands of a few

but information in the hands of many.

—John Naisbitt
Megatrends

Desktop GIS: What it is and what it does

Desktop GIS. If you've already heard of it, you may know it as an immensely powerful computer mapping system. It is that. But it is much more. It is a tool for managing information of any kind according to where it's located. With it you can keep track of where customers are, decide where to site businesses, manage sensitive wildlife habitats, optimize delivery routes, track the spread of infectious disease.

In this chapter, we'll show you some of the things that various kinds of organizations are already doing with desktop GIS. But before we do that, a few words about maps—those fundamental tools that help us communicate where we are, manage what's there, and figure out how to get someplace else.

Maps: who needs them?

You're flying home from a business trip. The woman sitting next to you has never been to your town, and you want to tell her how to get to the little café that serves the best omelettes in the world. How far do you get with "turn left on Main Street and go three blocks and turn right" before your acquaintance looks completely lost? So you grab a piece of paper and a pencil, and a few squiggles later you've identified a location, explained how to get there, and described the major landmarks she will pass along the way. And however crude or inaccurate your map, it does what every map does—it represents where objects are in the real world in relation to each other.

The urge to understand where we are is universal. The Babylonians recorded land ownership by drawing boundaries of parcels on clay tablets. The Mongolians painted the plans of their towns on their walls. The Chinese drew topographic maps on silk, using colored symbols to show locations of military installations in relation to streams, mountains, roads, and settlements.

Some people even mapped the winds and the seas. The Marshall Islanders made navigational charts with sticks for prevailing winds and wave patterns and shells for islands.

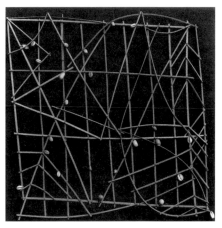

The Science Museum/
Science & Society
Picture Library

A nineteenth century version of a navigational chart. The South Sea islanders have probably been making these for thousands of years.

The Romans used paper maps to promote the growth of commerce in their rapidly expanding empire.

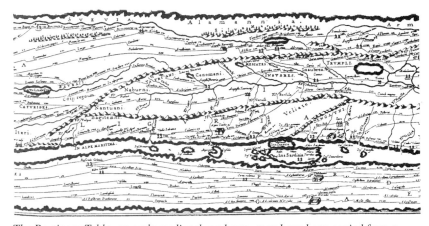

The Peutinger Table, an early medieval road map, may have been copied from a Roman map of the third century.

Whatever they were made from, wherever they were made, maps in the past shared two characteristics: they could only be made by skilled cartographers and they were static.

Mapping systems that run on desktop computers, however, give everyone the ability to make maps. And the maps they make can be changed in a flash, over and over again.

The diversity of mapping systems

The desktop mapping systems on the market today range from display-only systems like electronic atlases to full-featured geographic information systems (GIS). The dividing lines between one type of system and the next are not sharply defined. The systems do differ in a number of important ways: how they link geographic locations with information about those locations, the accuracy with which they specify geographic locations, the level of analysis they perform, and the way they present information as graphic drawings.

Electronic atlases, for instance, allow you to display pictures of geographic areas on your computer screen. They provide limited information about the geographic areas, and limited ability to alter the graphics. Without any tools for analyzing the information, these systems are most useful for providing graphics that can be used in presentations and reports.

Unlike electronic atlases, thematic mapping systems enable you to create graphic displays using information stored in a spreadsheet or database. These systems are especially useful for creating graphic presentations. Each map produced is based on a *theme,* such as population or income, and uses colors, patterns, shading, and symbols of various sizes to show the relative value of the information stored for that theme, at each geographic location.

ENVIRONMENTAL SYSTEMS RESEARCH INSTITUTE, INC.

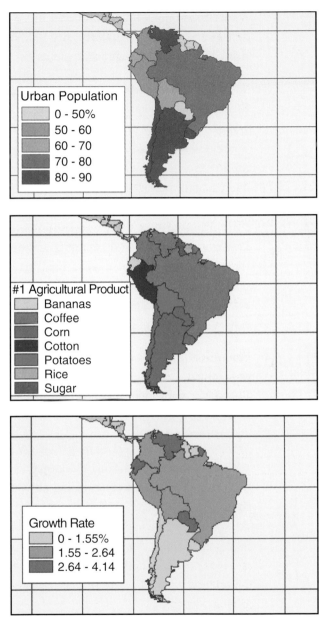

Population, agriculture, and income information are used to create these thematic maps.

Street-based mapping systems are more sophisticated than electronic atlases and thematic mappers. They link information to geographic locations. Street-based mapping systems can display address locations on street maps as points.

More sophisticated desktop mapping systems can import database or spreadsheet files or provide direct access to outside information sources. Some desktop mapping systems let you create and manage tabular information, use tabular information to create charts and graphs, and even analyze information statistically.

Desktop GIS puts it all together

Desktop GIS can do all these things and more. Desktop GIS combines all the capabilities of display-only, thematic, and street-based mapping systems along with the ability to analyze geographic locations and the information linked to those locations. Furthermore, you can either access information from the map or access the map from information.

And desktop GIS is dynamic. That means you can create maps that are not limited to a single moment in time. Simply update the information linked to a map and the map will automatically reflect those changes. You can do this quickly, without special training.

Desktop GIS lets you create map displays and maps for presentation simply by pointing and clicking. Desktop GIS lets you visualize and analyze information in new ways, revealing previously hidden relationships, patterns, and trends.

ENVIRONMENTAL SYSTEMS RESEARCH INSTITUTE, INC.

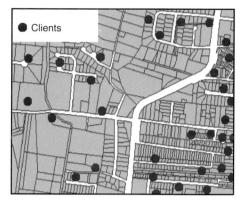

Visualizing customer locations is critical to businesses trying to make better marketing decisions.

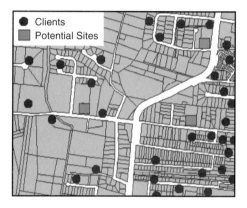

Analyzing location is key to making decisions about where to set up a business or service.

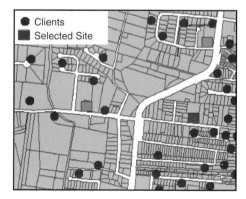

Presenting information as maps reveals relationships and patterns that may otherwise be hidden.

People in business, government, education, and natural resources are already using desktop GIS to analyze markets, manage parcels of land, conduct research, and protect natural resources. Desktop GIS can change the way you do business, whatever field you are in.

What people are doing with desktop GIS

Development that improves the land

Bob's development company has just purchased a beautiful hilly area covered with oak trees and fed by woodland streams. The draw for future residents is the area's "natural" feel, so he has to put in roads and houses without destroying the woods and streams. He also wants to avoid the steeper slopes of the hills. Local environmental regulations, which require him to avoid protected streams and sensitive bird habitats, complicate things even more. With desktop GIS, however, Bob can balance these conflicting needs.

He can quickly locate suitable areas by eliminating the unsuitable ones. And he can update, revise, and refine his information as often as he needs.

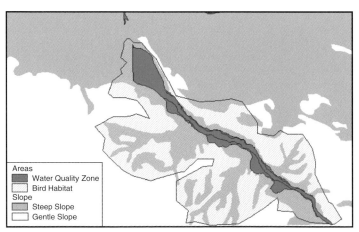

This map shows several areas (the white patches) that are suitable for development. They have gentle slopes and don't intrude on protected areas.

ENVIRONMENTAL SYSTEMS RESEARCH INSTITUTE, INC.

On the road again

Maintaining roads and highways and improving safety is a big job. John, a transportation planner, uses desktop GIS because it integrates information from a variety of sources.

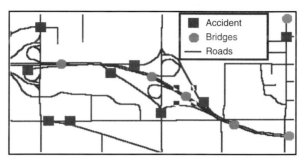

A transportation database might include roads, bridges, and locations of traffic accidents.

John's duties as a transportation planner include deciding which portions of which roads need to be repaired and when to repair them. Finding the road segments that require immediate action helps a planner decide where to allocate resources and when. It's not enough to know where roads are damaged. John must also know how badly they're damaged, how much traffic they carry, and when that traffic is heaviest. Only then does he know which crew to send out and where to send it.

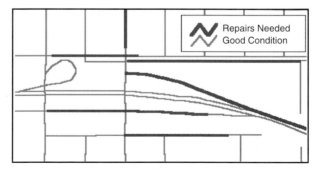

Road segments are displayed based on pavement condition and traffic volume. Segments in poor condition with heavy traffic become obvious on the map display.

John also analyzes traffic accidents to identify causes and plan corrective actions. This involves looking for places where accidents occur in clusters and then accessing information about those locations to try and determine possible causes.

John can retrieve information about accident sites, such as scanned images of accident reports, photos of accident sites, and tabular information about road conditions and traffic volumes.

The ability to link photos to these clusters can be particularly helpful in seeing contributing factors that might not otherwise be apparent, such as large trees or shrubs close to the road that block the view of a blind curve or intersection. Perhaps pruning the trees at these sites would improve conditions. Or maybe straightening a curve would help keep motorists on the road.

This old mall

Developers have been using GIS to choose sites for new shopping malls for some time. But what about when the mall is old and needs a facelift? Consider the case of Old Town Mall. Already in a good location, Old Town Mall is easy to get to. But it's almost thirty years old, and shows it. It has only two department stores, and they no longer pull in the customers. The smaller shops are also suffering.

Across the street from the mall is a forty-year-old open-air shopping center with one department store and a parking lot full of weeds and old newspapers. Joan, an energetic young developer, is thinking about buying both properties.

She has big plans: put all three department stores into the mall, make it two levels instead of one, and convert the open-air center across the street into a "power center" of large specialty discount stores selling electronics, appliances, clothing, toys, and the like. But she cannot act until she has a lot more information.

So Joan uses desktop GIS to find out who is likely to shop at the new mall, the demographics of the people who live in the area, and how the trade areas of competing malls compare with the new mall she is planning. First, she uses sales information linked to postal code areas to define the trade area for the new Old Town Mall.

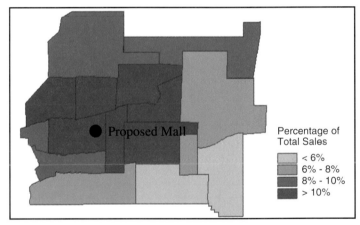

This map shows postal code areas ranked according to sales. Contiguous areas contributing 75 percent of total sales define the trade area for the mall.

Then she studies the demographic information to find out how many people live in the area and other things like how much money they make.

Trade Area ID	Population	Average Income
1	65504	35500

GIS calculates the population and average household income of the trade area and places these values in a table.

She then uses GIS to generate trade areas for each competing mall and compares them with the trade area for Old Town Mall.

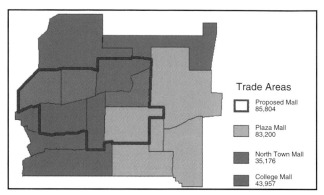

A map comparing the populations of trade areas shows a higher density in the proposed area.

Now Joan knows that more people live in the trade area for Old Town Mall than in any other trade area. But she needs to know how those numbers relate to income. Using the GIS software's ability to make charts, Joan can plot population and income categories for the trade areas of regional malls.

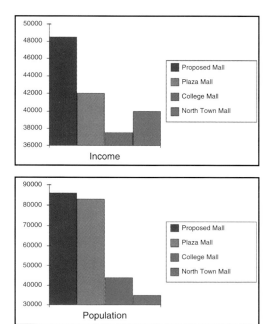

The bar charts demonstrate that the new mall's trade area demographics are as good as or better than the leading mall in every income category.

Since it looks to be a safe bet, Joan decides to buy the property and Old Town Mall is on its way to becoming a shopping center showcase.

The clearly compliant bank

Banks and financial institutions in the United States must comply with state and federal regulations and demonstrate that their lending practices do not discriminate. At the same time, they must show board members that they are profitable.

Midtown Bank is a perfect example. Henry, the branch manager, needs to show that Midtown Bank's loan distribution practices are not discriminatory. Until recently, that would have meant poring over stacks of spreadsheets and computer printouts, long after everyone else had gone home. But now, with the aid of desktop GIS, relationships, patterns, and trends become instantly apparent when the information is visualized on a color-coded map.

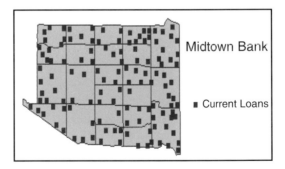

This map clearly shows a bank's lending patterns across neighborhoods.

With desktop GIS, Henry can see where Midtown's customers live and compare that information with demographics for the census tracts his bank serves.

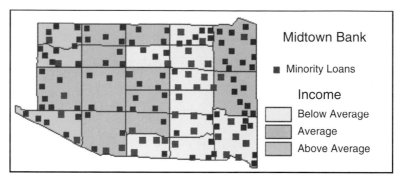

By overlaying demographic information, Henry can track and analyze income and minority loans.

This type of information can help Midtown Bank target prospective customers, compare its performance against that of the competition, and determine the best sites for expanding its services.

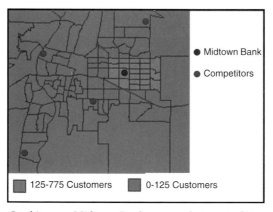

On this map, Midtown Bank can see that most of its customers live close to the bank. Seeing where its competitors are, on the same map, could help it decide where to target prospective customers.

Midtown Bank is discovering that desktop GIS brings new insights into many phases of operating and managing a financial institution.

ENVIRONMENTAL SYSTEMS RESEARCH INSTITUTE, INC.

The fire's aftermath

Fire fighting is hard work. But after the fire is out, even more work remains. Blackened structures need to be cleared away and new ones built. Polluted streams need to be cleaned up. Utility lines need to be repaired. Rachel, who works in the county planning department, uses desktop GIS to evaluate everything from which building codes may have contributed to the fire's spread to which routes will be best for emergency teams and cleanup crews.

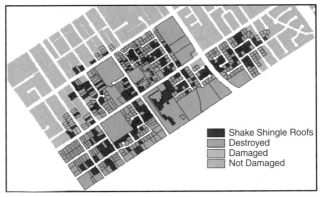

This map shows which houses were destroyed and which survived but were damaged. Homes with wood shake roofs were identified as major culprits in the fire's spread.

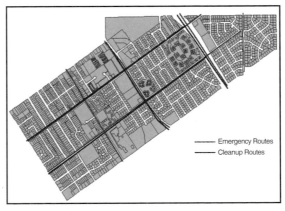

Rachel combines parcel and street maps to identify the best routes for getting emergency equipment into an area and debris out of it.

After the fire, other effects need to be considered. A fire in hilly country commonly results in landslides. The county soils department uses GIS to provide Rachel with information about landslide-prone areas so they can be reinforced before rebuilding, or avoided completely.

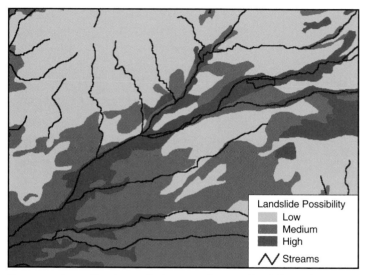

By comparing soils, slopes, drainage, and vegetation information, areas most prone to landslides can be identified.

GIS can supply this kind of information and may be the best tool for avoiding disasters in the future.

Chemicals where you need them

Ted has a hard job. He's a farmer. He not only has to battle the weather and various pests, but he has to do it without endangering the environment and without losing money.

One way Ted can save money and still protect the environment is by applying chemicals such as fertilizers and pesticides to fields only where they are needed instead of applying them uniformly. But this requires more detailed information about the health of each field, and the information comes from a variety of sources, such as field work, maps, and photographs.

GIS's ability to link descriptive information and photographs with maps makes it a natural solution. Ted combines infrared images with maps and information about soil types and nutrient status. This helps him identify problem areas, evaluate crop stress, and target specific areas for fertilizer applications.

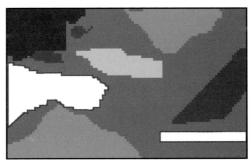

This crop stress map shows the lighter areas as low in nutrients; they should be targeted for fertilizer applications.

Since the process of identifying problem areas is visual, maps are perfect tools. Ted need only draw a perimeter around a problem area on a map to identify it. Then, if a ground sample shows an unusually low yield or the presence of a parasite, he knows where to target fertilizers or pesticides.

Providing the care where it's needed

Whenever Theresa has an asthma attack, she often ends up in the emergency room of the local hospital. But emergency room care is expensive, and many conditions such as asthma can be successfully treated in an outpatient setting, especially when care is administered early.

The problem comes when the type of outpatient care needed is not accessible to patients. Brookside Hospital decided to use desktop GIS to solve this problem. With GIS's mapping capabilities, and access to emergency room records, hospital staff were able to map the locations of thousands of patients who had visited their emergency room. They also mapped the locations of community clinics and primary care physicians.

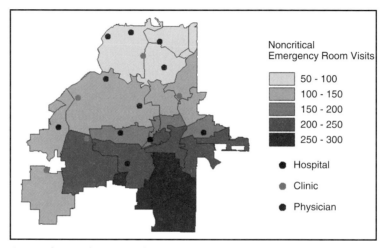

A map showing locations of patients, clinics, and primary care physicians shows the gaps in access to primary care.

Using these maps, the Brookside staff were able to better focus the available primary care resources, and determine where more family, primary care, and general practice physicians were needed.

ENVIRONMENTAL SYSTEMS RESEARCH INSTITUTE, INC.

Whose woods are these?

People need space to live in. Animals need space to live in. Sometimes it's the same space. Larry works for a wildlife organization that has begun to use desktop GIS to identify where wild species could live without using human space.

The first step is generating maps of wildlife habitats, one map per species.

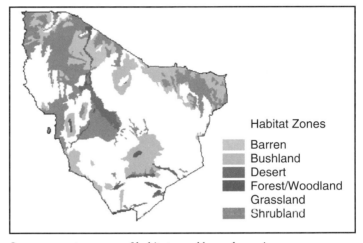

Larry generates a map of habitats used by each species.

Larry must consider each species separately, because each one ranges differently. Some nest in one place and forage in another. Some migrate and need corridors to get from winter to summer homes and back. Some range widely and the whole of their range must be preserved.

Larry then combines all of the species maps to determine which areas are used by the greatest number of species.

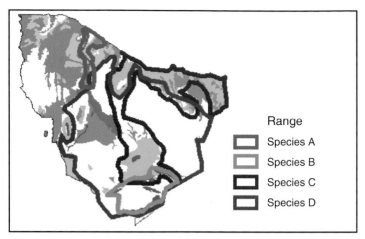

When habitat maps of all species are combined, those areas used by the largest number of sensitive species become obvious.

When Larry combines these maps with public land maps, he sees immediately what needs to be done.

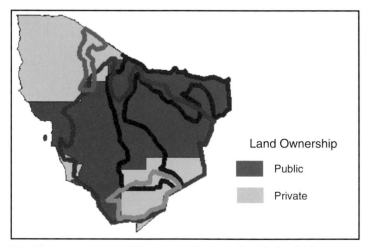

Combining maps of wildlife areas with those of public lands shows the extent to which species are already being protected on public lands. It also shows areas not currently in public ownership and in need of protection.

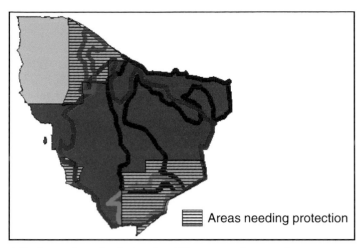

A final map shows lands that should be protected to meet long-term conservation needs of sensitive species.

These results give Larry's organization what it needs to work with governments, planning councils, and the private sector to make a place where all animals can have a future.

What you can do with desktop GIS

You don't have to be a farmer or a wildlife manager. You don't have to be a developer or a traffic engineer. If what you do involves managing information, and that information can be linked to geographic locations, then GIS can help you organize that information in new ways so that you can make new discoveries and get more out of the information you have. The possibilities are endless.

So read on. The next chapter gives you the basic concepts you need to understand the technology. And we continue to address the topic of what you can do with desktop GIS throughout the first part of this book.

This is how it works

The most incomprehensible thing about the world is that it is comprehensible.

—Albert Einstein

This is how it works

Desktop GIS represents the real world on a computer similar to the way maps represent the world on paper. But desktop GIS has power and flexibility that paper maps lack. This chapter will introduce you to the basic concepts of desktop GIS—what it has in common with maps and what's different.

To help you understand how desktop GIS works, we'll break it down into the conceptual pieces, and take them one piece at a time. We'll start with basic ideas, like how maps convey information about places and how scale influences the size of what appears on the map. Then we'll advance to the principles governing how a desktop GIS works, like how it stores and links information and where it gets its power and flexibility.

What you see on a map

Map features represent objects in the real world

Maps are graphic representations of the real world. Since the real world is infinitely more detailed than a map can be, we say that maps give us a generalized view of the real world. Natural objects, such as mountains, rivers, and valleys as well as man-made objects such as cities, roads, and buildings can all be represented on maps.

The objects represented on maps, whether natural or man-made, are called *map features,* or simply *features.* Looking at a map of downtown Portland (on the next page), we can identify some of its features.

ENVIRONMENTAL SYSTEMS RESEARCH INSTITUTE, INC.

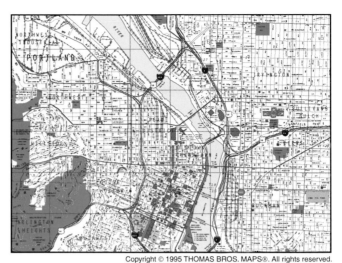

The features representing downtown Portland are shown on this map.

Each map feature has a location, a representative shape, and a symbol that represents one or more of its characteristics. The large blue area represents the Willamette River. The white areas represent land. Some large green areas stand out—these represent parks; a yellow area says "Lloyd Center"; a gray area says "Lonefir Cemetery." Freeways are marked by dark red lines, and smaller roads are marked by thin gray or black lines.

The relationships between locations

The locations of map features reflect more or less accurately their locations on the earth's surface. Because the earth is a sphere and maps are flat, there is necessarily some distortion in the locations of features on maps. We'll discuss the distortion found on maps, how to control it, and the methods used for recording locations of features in chapter 5.

Because features on maps are organized according to relative position or location, maps are particularly good for showing the relationships between feature locations. These relationships, called *spatial relationships,* are important because understanding them helps us solve problems. For example, in order to plan a delivery route, you need to know which streets connect, where they cross highways, and which parts of town they pass through. Selecting a site for new picnic grounds might involve finding areas near the river, adjacent to office buildings, and where there are plenty of trees.

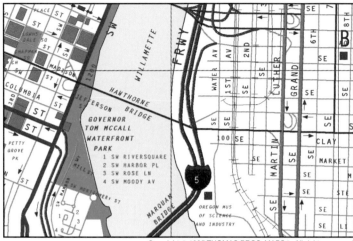

The relationships between feature locations are visible on a map. You can see where highways connect and which ones cross the railroad tracks or the river; you can see which buildings are near the river and also near parks.

Map features have distinct shapes

To represent real-world objects, maps use three basic shapes—points, lines, and areas (commonly referred to as points, lines, and polygons in a GIS). Any object can be represented using one of these shapes.

Points represent objects that have discrete locations and are too small to be depicted as polygons. On the Portland map, these include schools, churches, train stations, fire stations, and other buildings such as museums and the Department of Motor Vehicles.

ENVIRONMENTAL SYSTEMS RESEARCH INSTITUTE, INC.

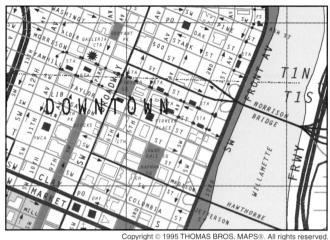

Small buildings, such as train stations, post offices, the YWCA, and the Department of Motor Vehicles, are represented as point features on the Portland map.

Lines represent objects that have length but are too narrow to be depicted as polygons. On the Portland map, lines represent freeways and roads of all kinds, railroads, bridges, and creeks.

Freeways, bridges, and roads are prominent line features on the Portland map.

Areas represent objects too large to be depicted as points or lines. The Willamette River, parks of all kinds, Portland State University, the Lloyd Center, Buckman Field, and Lonefir Cemetery are all shown as areas on the Portland map.

The river, parks, and large buildings are area features on the Portland map.

Symbols identify or characterize map features

Shapes alone do not give you enough information. So, maps use graphic symbols to help identify features and provide information about them. There are symbols for points, symbols for lines, and symbols for areas. Symbols for points often look like the features they identify. For example, the symbol for a school may be a little red schoolhouse and the symbol for an airport may be a small plane.

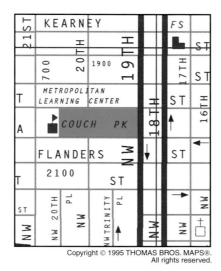

Schools, churches, and fire stations all have distinct symbols on the Portland map.

Line symbols include thick or thin lines, solid or broken lines, and may come in colors.

ENVIRONMENTAL SYSTEMS RESEARCH INSTITUTE, INC.

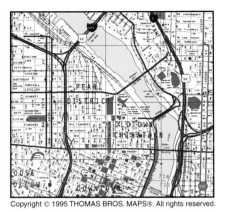

Red double lines make the freeways stand out and differentiate them from other roads; thick, black lines differentiate major roads from lesser roads (thin, gray lines).

Area symbols include the colors and patterns used to fill in areas. Some colors have a natural connection to the objects they represent, such as blue for water and green for parks and forested areas, while others do not.

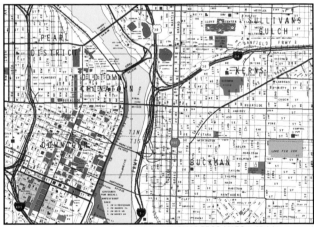

The river is shaded blue, parks are green, the university is pink, the Lloyd Center is yellow, and the cemetery is gray.

Sometimes symbols don't give enough information and so text labels are added to help identify features. For example, you know that the blue area on the Portland map is a river, but you may not know which river; you may know that the green areas are parks, but you need labels to identify them by name.

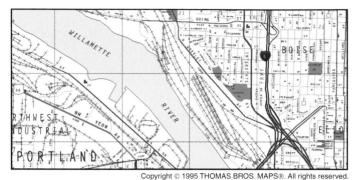

Features of all kinds need labels to identify them.

Map scale determines the size and shape of features

Most features can be represented as more than one shape. The *scale* of a map tells how the size of the map features compares with the size of the geographic objects they represent. The larger the map scale, the bigger the map features will appear. Depending on the map scale, a feature such as a city can appear as a point or as an area, and a feature such as a river can appear as a line or as an area. For example, the Willamette River is represented as an area on the Portland map, but on a map of Oregon State it appears as a line. The city of Portland covers the entire area shown on the downtown Portland map, but the same area appears as a single point on the state map. The buildings that are points on the Portland map would appear as areas on a larger-scale land use map.

Lines depict roads and rivers alike on this smaller-scale state map; cities are depicted as points.

ENVIRONMENTAL SYSTEMS RESEARCH INSTITUTE, INC.

Buildings appear as areas on this land use map.

What's different about desktop GIS

Desktop GIS links features with lists of attributes

On paper maps, each color, pattern, picture, or label gives you information about the features. But, the amount of information you can get from a paper map is limited to what is shown.

With desktop GIS, you can get an almost unlimited amount of information about what you see on a map. Desktop GIS stores all the information about map features in a GIS database and links the features on the map to the information about them. This means that you can access all the information about a feature by simply clicking on it.

The information that a desktop GIS stores about map features is referred to as *attribute* information, or *attributes*. The attributes of a river, for example, might include its name, length, average depth, rate of flow, water quality, how many dams are on it, and how many bridges cross it. The attributes for a map feature that represents a shopping mall might include the name of the mall, its type, size, the names of its anchor stores, a list of tenants, and the number of available spaces.

Desktop GIS formats attributes in rows and columns, and stores them as *tables*. Each column stores a different attribute and each row relates to a single feature.

Street ID	Length	Surface Material	Resurface Date	Speed Limit	Number of Lanes	Avg. Daily Traffic
66	5.4	asphalt	5/85	40	4	3200
69	9.4	asphalt	6/88	50	4	4000
99	25.3	concrete	7/91	55	6	7900

The attributes of all the streets on a map can be stored in a table like this.

The link between map features and their attributes is the basic principle behind how a desktop GIS works, and is the source of its power. Once the map features and attributes are linked, you can access the attributes for any map feature or locate any feature from its attributes in a table.

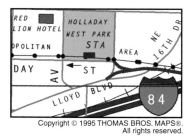

Park Name	Area (acres)	Admin.	Type Code
Waterfront	26	city	2
Holladay	2	city	1
Irvin			

Type	Sq. Feet	# of Floors	Year Built
office	12000	5	1989
office	50000	15	1980
hotel	110000	22	1991

Street ID	Length	Surface Material	Resurface Date	Speed Limit	Number of Lanes	Avg. Daily Traffic
81	1.4	asphalt	6/94	35	2	200
82	19.4	asphalt	10/89	40	4	2400
84	34.5	concrete	2/90	55	8	10100

Pointing at any feature on the Portland map displays the list of attributes linked to that feature. First you see the attributes for a park, then a building, then a road.

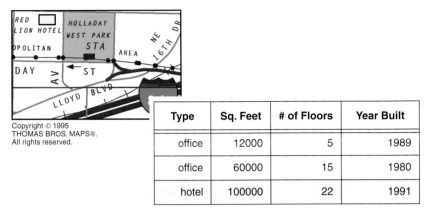

Type	Sq. Feet	# of Floors	Year Built
office	12000	5	1989
office	60000	15	1980
hotel	100000	22	1991

From the table of attributes, you can access the building feature linked to any attribute you choose.

Desktop GIS displays features based on their attributes

Not only can GIS access features from an attribute table and access attributes from a map, it can also display features based on any attribute in the table. Take a look at how this works using the streets of downtown Portland. Suppose you need to move a modular office building across town to the Lloyd Center. You'll need to know which streets are wide enough to carry the extra-wide load. The street features are stored in a desktop GIS and linked to a table of attributes. Recall that one of the attributes is the number of street lanes. You display the streets based on the number of lanes for each street so you can find the widest streets. Then you can trace a path across town that follows these streets.

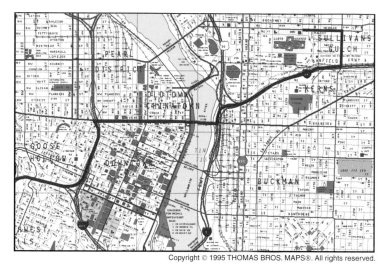

This path follows the widest streets across town.

But wait a minute. Another important variable to consider is the type of traffic you'll encounter. So you display the streets again, this time based on a different attribute, average daily traffic. Now you can trace a different path, one that follows streets with the lowest daily traffic flow.

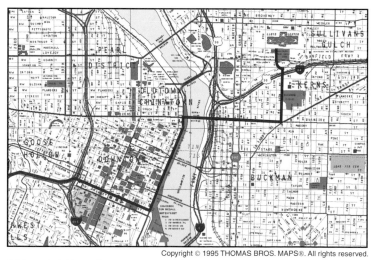

This path follows the streets with the least traffic.

Now you might ask, "What if I want to find streets based on both attributes?" That is, you want to find wide streets that have low traffic flow. Because features and attributes are linked, this is easy to do with desktop GIS. The GIS can locate the features you want based on any number of attributes and display them on a map. In the next chapter, we'll discuss the details of how this works.

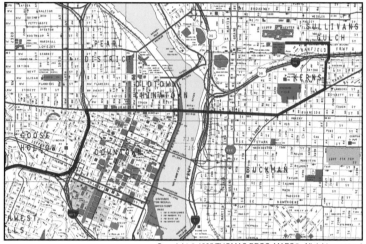

This path follows the widest streets with the least traffic.

The link between features and attributes is dynamic

Because the link between features and attributes is a two-way relationship, changing an attribute in the table automatically results in a change on the map. Here's a hypothetical situation to demonstrate how it works. Say that Fremont Street, located in the northeast portion of downtown Portland, is a major road between Highway 99E and the eastern edge of town. But, west of Highway 99E, it's a minor road.

The change in line symbols from thin to thick shows that Fremont Street changes from a two-lane to a four-lane road as it crosses Highway 99E.

Suppose that the highway department has decided to widen the western portion of Fremont Street, from its current two lanes to four lanes, so both portions will have the same number of lanes. You work in the GIS division, and it's your job to update the attribute table to reflect this change. Since the map symbol for a four-lane road is different than for a two-lane road, the symbol for the western portion of Fremont Street should change once you update the table.

Street ID	Length	Surface Material	Resurface Date	Speed Limit	Number of Lanes	Avg. Daily Traffic
66	5.4	asphalt	5/85	40	4	3200
69	9.4	asphalt	6/88	50	4	4000
99	25.3	concrete	7/91	55	6	7900

In the attribute table, you change the number of lanes from two to four for the western portion of Fremont Street.

The next time you display the road features, the western portion of Fremont Street appears with the same line symbol as the portion east of Highway 99E.

This simple link between features and attributes makes desktop GIS a truly dynamic system.

How desktop GIS manages features and attributes

Themes link features with their attributes

Desktop GIS links sets of features and their attributes and manages them together in units called *themes.* A theme contains a set of related features, such as roads, streams, parcels, or wildlife habitat areas, along with the attributes for those features. Take, for example, the Portland map. It contains many themes. All the interstate freeways could make up one theme, and all railroads, another. City streets might be a separate theme. Parks, buildings, and waterways are examples of other themes.

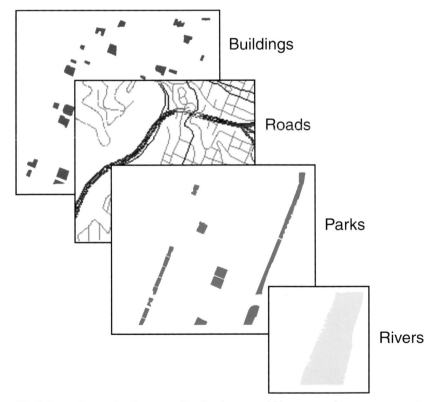

Buildings

Roads

Parks

Rivers

The information on the downtown Portland map could be managed as many separate themes.

Themes are made up of features with a set of common attributes. For example, all the roads have lanes, pavement type, and a route or street name. On the other hand, the railroads have a different set of characteristics in common, such as whether they are main line or branch line railroads, the type of usage they get, and the volume of traffic they bear, measured in tons.

Sometimes features that share common attributes are placed in separate themes for convenience. For example, if you work with and display freeways separately from other roads most of the time, you can keep them in a separate theme.

Collections of themes form a GIS database

All the themes for a geographic area taken together make up a *GIS database*. You can use the themes in a GIS database to analyze multiple situations and solve multiple problems. To determine the best routes for a pickup and delivery service, you could use the freeway and road themes. To plan a tour of the city, you might use the themes for roads, buildings, parks, and points of interest.

The design of a GIS database is strong because it's flexible. You can add new themes to a GIS database or delete old ones; you can separate themes to create more themes, or combine themes if they have common characteristics. What you want to do with a GIS database, and what information you need, will determine the best design for you.

ENVIRONMENTAL SYSTEMS RESEARCH INSTITUTE, INC.

Asking questions;
getting answers

He who asks a question is a fool for five

minutes; he who does not ask a question

remains a fool forever.

—Chinese proverb

Asking questions; getting answers

At first glance, a GIS map display on a computer screen looks like any other map. The solid black lines are roads, the thin blue ones are rivers; the circles are cities, the small triangles are mountain peaks. But with a GIS map display you can get detailed information about each feature; with GIS you can find features based on their attributes and analyze feature locations to uncover relationships between them.

Finding attributes by selecting features

Suppose that you're a real estate agent. You need information for a client about a piece of property. With desktop GIS, you're only a click or two away from displaying a table of the property's attributes, a photograph of the property, a legal description of it, real estate figures about it, and even a video of the inside of the house located there. Here's how.

Selecting features by pointing

You recall from chapter 2 that desktop GIS links features with their attributes and stores them as a single row in an attribute table. When your clients want to know the attributes of a particular property, like when a house was built or how many bedrooms it has, you need only point to the property on the computer map with a mouse to see its attributes.

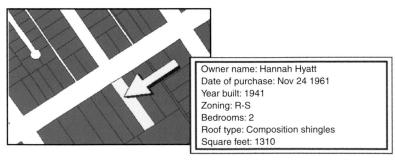

Owner name: Hannah Hyatt
Date of purchase: Nov 24 1961
Year built: 1941
Zoning: R-S
Bedrooms: 2
Roof type: Composition shingles
Square feet: 1310

Pointing at a property on this map displays its attributes.

ENVIRONMENTAL SYSTEMS RESEARCH INSTITUTE, INC.

Now you know who currently owns the property, when it was purchased, how it's been improved, how many bedrooms the house has, and how big it is. Do you need to know more? If so, desktop GIS enables you to access additional information as long as that information is linked to features on a map display.

Legal description:
2ND PREL MAP ADD NO 3 PTN LOT 5
3/4 BLK 1 COM ON SLYL1 FERN AVE
S 56 DEG 20MINW198FT FROM MOST
NLY COR SD LOT TH N 56 DEG 20

Pointing at the property this time displays its legal description and a picture of the house located there.

Selecting features by drawing shapes

Sometimes you need to work with more than one feature at a time, like when your clients want information about all the homes for sale in a neighborhood. Selecting many features individually by pointing is tedious and time consuming. So, to list and compare information about a group of features, you select them by drawing a line through them or a shape around them.

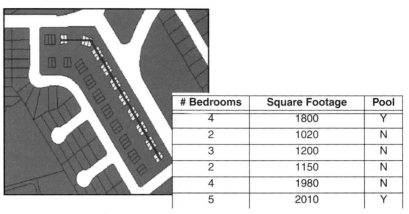

# Bedrooms	Square Footage	Pool
4	1800	Y
2	1020	N
3	1200	N
2	1150	N
4	1980	N
5	2010	Y

Drawing a line selects all the houses the line passes through along with their attributes.

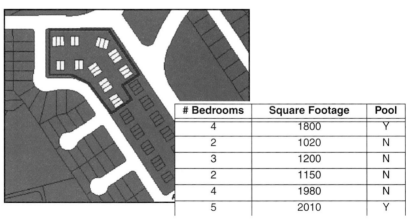

# Bedrooms	Square Footage	Pool
4	1800	Y
2	1020	N
3	1200	N
2	1150	N
4	1980	N
5	2010	Y

Drawing an irregular shape selects all the houses inside the shape along with their attributes.

Finding features by selecting attributes

Pointing at rows

You've seen some ways to select features directly from the map display. But, since features and attributes are linked, you can also select features indirectly by selecting their attributes. By pointing at a row in an attribute table, you can select it, along with the feature it's linked to. You can select one row or as many rows as you like. Features linked to the rows you select are highlighted on the map display. This can be a powerful way to find features while looking over a table of their attributes.

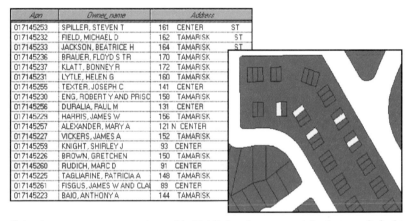

Apn	Owner_name	Address		
017145253	SPILLER, STEVEN T	161	CENTER	ST
017145232	FIELD, MICHAEL D	162	TAMARISK	ST
017145233	JACKSON, BEATRICE H	164	TAMARISK	ST
017145236	BRAUER, FLOYD S TR	170	TAMARISK	
017145237	KLATT, BONNEY R	172	TAMARISK	
017145231	LYTLE, HELEN G	160	TAMARISK	
017145255	TEXTER, JOSEPH C	141	CENTER	
017145230	ENG, ROBERT Y AND PRISC	158	TAMARISK	
017145256	DURALIA, PAUL M	131	CENTER	
017145229	HARRIS, JAMES W	156	TAMARISK	
017145257	ALEXANDER, MARY A	121 N	CENTER	
017145227	VICKERS, JAMES A	152	TAMARISK	
017145259	KNIGHT, SHIRLEY J	93	CENTER	
017145226	BROWN, GRETCHEN	150	TAMARISK	
017145260	RUDICH, MARC D	91	CENTER	
017145225	TAGLIARINE, PATRICIA A	148	TAMARISK	
017145261	FISGUS, JAMES W AND CLA	89	CENTER	
017145223	BAIO, ANTHONY A	144	TAMARISK	

Selecting one or more rows in a table highlights the rows and the features linked to them.

Entering text

Looking for a feature on a paper map can be challenging if the feature is small. With GIS map displays, finding a feature is easy no matter how small it is, because you can find it by requesting any of its attributes. Suppose your client wants to purchase a commercial property. He's seen a For Sale sign in front of a vacant building. He jots down the address and brings it to you on the back of a crumpled business card. You enter the address as a request. Desktop GIS locates the feature linked to the address you enter and shows it to you by highlighting it on the map.

Find: **504 West Main**

By entering a street address, you can find the property at that address.

You could find the same property by entering a different attribute. Suppose your client had written down the building's name, "Bradshaw Building," instead. You can simply enter that name, and the first building found that matches your request will be highlighted on your screen.

Finding features that meet your criteria

You can find one or more features by requesting them based on an attribute they share. For example, you can ask for buildings of a certain type, buildings built in a certain year, or buildings with a certain square footage. All the buildings that match your request are selected and highlighted in both the attribute table and on the map display.

You can find features based on more than one criterion, too. A couple has asked you to look for a house with a tile roof (for fire safety) and five bedrooms (they have four kids). You write a request that asks for houses having both "tile roofs" and "five bedrooms."

ENVIRONMENTAL SYSTEMS RESEARCH INSTITUTE, INC.

The desktop GIS finds and selects all the houses with both of these attributes from all the other houses in the map theme. (Recall that a theme stores all the features of a particular type, such as houses, along with their attributes.) The desktop GIS highlights the selected houses on the map and in the table containing their attributes.

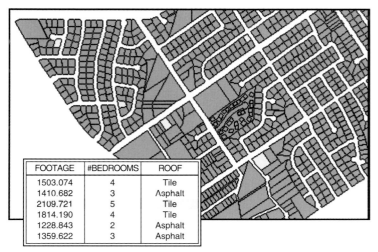

FOOTAGE	#BEDROOMS	ROOF
1503.074	4	Tile
1410.682	3	Asphalt
2109.721	5	Tile
1814.190	4	Tile
1228.843	2	Asphalt
1359.622	3	Asphalt

This map highlights parcels with houses that have tile roofs and five bedrooms. The attributes linked to the selected features are also highlighted in the table. (Only part of the attribute table is currently visible.)

Performing operations on selected features and their attributes

Once you've selected a group of features, you can perform any operation on them as a separate group, apart from all the other features. For example, you can have the GIS zoom in to see selected features fill the center of your screen, or perform statistical operations on any numeric attribute, or use statistics to summarize the information in a table, or even create a chart comparing attributes of the selected features.

Your clients want to know prices and number of square feet for the selected houses. You have a column for price and one for square footage in the attribute table. The GIS can compute the price per square foot and place the results in a new column you've added to the table. Now that you know the average price per square foot for houses in the selected group, you can combine all the information to display only those houses that meet all your clients' criteria.

This map shows parcels with houses that have tile roofs, five bedrooms, and a price of less than $100 per square foot.

It's the relationships that matter

In chapter 2, we examined the features on the downtown Portland map and the relationships between them—freeways crossing the river, roads intersecting other roads, neighborhoods sharing common boundaries. Some buildings were near the river, others were near the freeways. The parks were on the west side of town. Some features even shared the same geographic space (the Lloyd Center contained several other features).

The orientation of certain features to those around them may be important if you need to know such things as which schools are within a certain distance of properties you're considering, or whether a specific property is next to a commercial zone, inside this fire district or that water district or that flood zone, or underlain by clay soils or a fault zone, or overlapping an endangered species habitat.

When you use desktop GIS to analyze these relationships, you are performing *spatial analysis.* "Spatial" refers to the way information is organized on maps, that is, according to its relative position or location on the surface of the earth. Analyzing spatial relationships is something a desktop GIS is very good at.

How far is it?

One way of analyzing the locations of features is by measuring the distance between them and other features in the area around them. The information you get by clicking on a piece of property will probably show you the dimensions of a lot, but probably not how far the house is from the shopping mall, the distance to the nearest airport, or the closest high school.

Measuring distance on a GIS map is easy. When you enter two points with the mouse that define the distance you want to measure, the desktop GIS calculates and reports the distance between them in any units you choose (e.g., feet, meters, miles, kilometers).

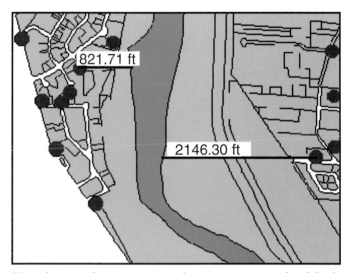

Here, distances from properties to the river are measured and displayed; properties within 1,000 feet of the river have higher insurance rates.

How big is it?

Another way of analyzing the locations of features is by measuring the area around them. Perhaps you need to know the size of a corridor between a property and the soon-to-be-built freeway or the area surrounding a proposed airport that buffers it from the surrounding residential properties. Desktop GIS calculates area as easily as distance. Just draw any shape on a map and the GIS calculates the area inside that shape.

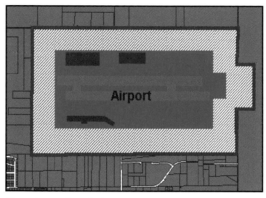

63758328 sq ft

Drawing a closed shape around the airport lets you calculate the size of the area that separates it from surrounding residential properties.

Finding the features nearby

A GIS can find features located within a certain distance of other features you specify. For example, the decision to buy or develop a certain property could depend on its proximity to surrounding features. Parents might want their house close to a particular school. A developer might purchase a property only if it's far enough away from the floodplain. An investor might be interested in a shopping center only if it's near major highways.

Say your clients with the four kids now want you to find all the houses for sale within half a mile of a particular school, so their children can walk to it. The houses and schools are in separate map themes, but you can display both themes at the same time.

The house and school themes are both displayed.

Then you can select the particular school and find all the houses within a half-mile radius of it. Since you're only interested in the ones that are for sale, you can select only those.

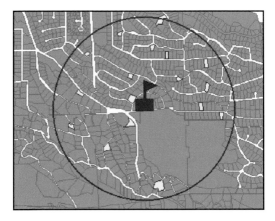

Only houses for sale within half a mile of the school are highlighted on the map.

Finding the features inside

A GIS can find points, lines, and areas that are enclosed within other area features. You might want to find all the fire hydrants in a subdivision containing undeveloped lots. This may be important to clients who want to build their own homes because lots far from fire hydrants may require that you pay to have a hydrant put in or that you install an expensive sprinkler system in your home.

The subdivisions and fire hydrants are stored in separate themes in the GIS database, but this is not a problem because desktop GIS allows you to find features in one theme contained within the boundary of a feature in another theme. First you display the subdivision and hydrant themes together. Then you select the subdivision of interest from the subdivision theme. The desktop GIS finds all the fire hydrants located within the selected subdivision, highlights them on the map, and selects them in the GIS database.

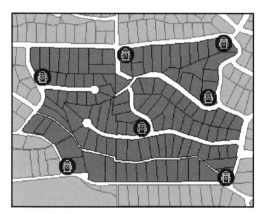

This map shows fire hydrants located within the selected subdivision.

Now you notice that scenic Henderson Creek runs through certain properties in the subdivision, making them highly desirable to people who want to build close to the river. The GIS finds all the properties Henderson Creek crosses and you print out a list.

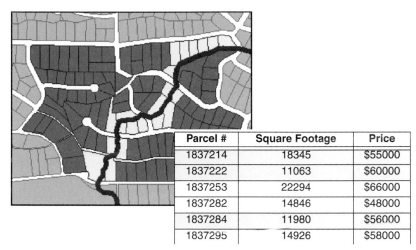

Parcel #	Square Footage	Price
1837214	18345	$55000
1837222	11063	$60000
1837253	22294	$66000
1837282	14846	$48000
1837284	11980	$56000
1837295	14926	$58000

Properties crossed by Henderson Creek are highlighted and selected.

Working with properties is not the exclusive domain of real estate agents. Local governments need land use or land ownership information too. Whenever there is a zoning change, for example, the city must notify the owners of the affected properties. Using the boundary of the new zone, the desktop GIS can identify all affected properties and provide the names and addresses of their owners.

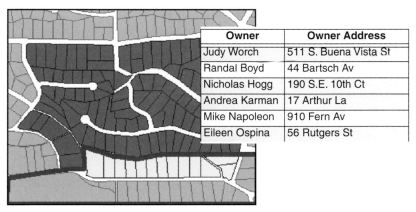

Owner	Owner Address
Judy Worch	511 S. Buena Vista St
Randal Boyd	44 Bartsch Av
Nicholas Hogg	190 S.E. 10th Ct
Andrea Karman	17 Arthur La
Mike Napoleon	910 Fern Av
Eileen Ospina	56 Rutgers St

The desktop GIS selects the properties that would be affected by a zoning change and displays their owners' names and addresses.

Finding the features next to other features

Desktop GIS also locates features that are next to other features you select. This type of analysis is important when selecting sites for businesses. For example, your client wants to purchase a space where she can locate a new store and expand into neighboring spaces if she needs more room. So she wants to know the square footage of the spaces adjacent to the one she wants to buy—those that share walls with her space. Desktop GIS finds adjacent properties, highlights them on the map, and selects them in the GIS database.

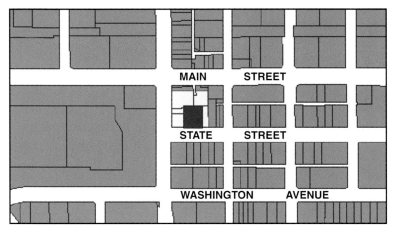

Properties adjacent to the selected business (red) are selected and highlighted on the map (yellow).

Finding where features share space

A property is a certain piece of land. The property is located in a certain land use area and building zone. It is also in a certain school district, fire district, and emergency services district. Not only that, the land is a habitat for certain plants and animals. It is made up of certain soils and certain geology layers, and below it, there may even be groundwater.

A GIS database stores all of these objects as map features in separate map themes, one for properties, one for habitats, one for geology, and so forth. Anytime you want to know which features occupy the same geographic space, you can display the themes on top of one another. This allows you to visually inspect which features share this same piece of the earth.

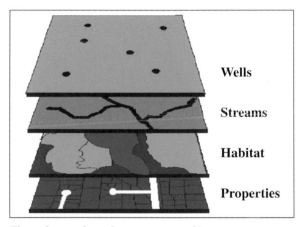

These themes share the same geographic space.

A desktop GIS uses the location information associated with each feature to find the exact intersection of the features, no matter what theme they are in. This process of finding features that share common space is a powerful tool called *spatial overlay.*

Suppose a city needs a suitable site for a landfill. The engineers would have a list of criteria to start with. For example, a landfill must be located on land with a certain type of soil that is rich in clay, to prevent materials buried in the landfill from seeping into the groundwater below.

The landfill must also be located away from airports. This is because birds visiting the landfill might collide with airplanes, causing potentially catastrophic accidents. To prevent the movement of material once it's in the ground, a landfill should not be located in an earthquake fault zone or in a flood zone. To meet these criteria, the engineers would use GIS to find those areas that are not too close to airports and not within a fault or flood zone.

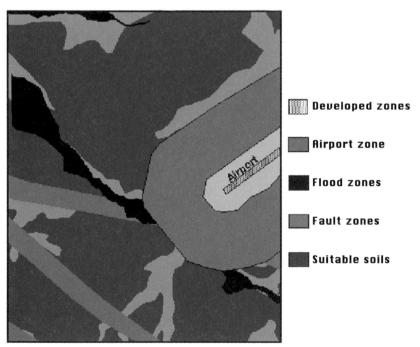

Developed zones

Airport zone

Flood zones

Fault zones

Suitable soils

This map shows areas where soils are acceptable, outside the 5-mile radius around the airport, and outside the flood and fault zones.

And since nobody wants a landfill near their house, the engineers will have to find the areas that do not overlap with residential streets. By selecting some areas and avoiding others, they find all the areas that meet all the criteria for a landfill.

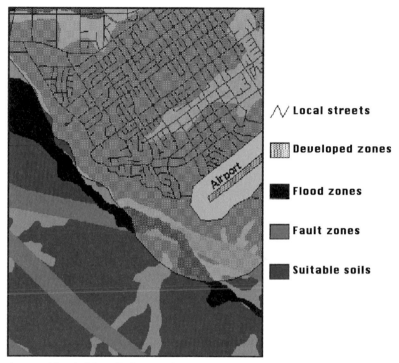

Local streets

Developed zones

Flood zones

Fault zones

Suitable soils

Areas with suitable soils that are far enough from airports, flood zones, fault zones, and residential streets are potentially good sites for a landfill.

You provide the questions; desktop GIS provides the answers

As you work more with desktop GIS, you'll find that it often leads you to ask new questions. Sometimes the answers present new information, even surprises. But no matter how many or how complex the questions, desktop GIS gives you the tools for understanding and analyzing the information so you can get the answers you need.

ENVIRONMENTAL SYSTEMS RESEARCH INSTITUTE, INC.

Making information presentable

To envision information—and what bright and splendid visions can result—is to work at the intersection of image, word, number, art.

—Edward R. Tufte
Envisioning Information

Making information presentable

Now you know how to ask questions and get answers from a desktop GIS. But how do you present the answers you get? Often it's not enough to have the answers yourself. You may have to present the information to others in a way that convinces them that the information you've derived supports your point of view.

Desktop GIS can help because it provides a wide variety of tools for presenting information as quality graphic presentations. These presentations may include maps, charts, and tables, along with graphics you import from other programs or even graphics you draw yourself. The presentations you create can be output to a printer to produce hard copy, or displayed on your computer's screen.

To make information presentable, you need to know your audience. Then you need to decide which information to include in your presentation. If your presentation includes maps, then how much information should be included on each map, and how should the information be organized and displayed? Should you use charts or tables instead of maps or in addition to maps? What other graphics could enhance your presentation?

We'll explore each of these topics separately, and then show you how easy it is to create attractive presentations that accomplish your objectives.

Knowing your audience

Before you sit down to create a graphic presentation that other people will see, ask yourself who they are and how much they already know. Are they a general audience or a group with specific knowledge of the topic you're presenting? Say your presentation includes a map. Knowing who is going to see the map will determine the colors and symbols you choose, the amount of detail you show, and how you organize the information. Only when you know who your audience is can you create the presentation that best conveys your message.

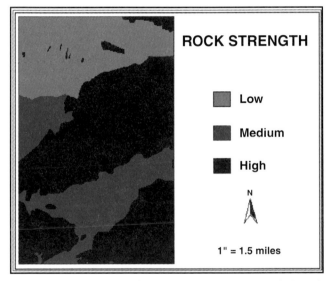

This map was designed for a public meeting. It uses symbols and text that anyone can understand.

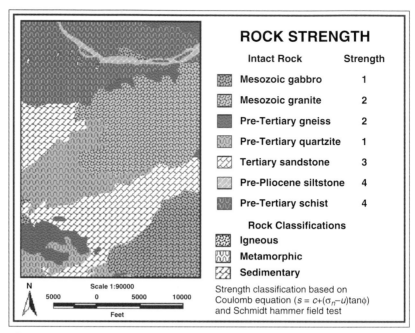

This map was designed for scientists. It presents similar information but with more detail, and uses symbols and text that a person with a technical background understands.

Knowing what information to present

You have a GIS database full of information, and you can access any of it. But, how much of this information should you present on a map? Too much information will confound and overwhelm the audience, and won't get your message across. If you get reactions like these, you've probably botched it: "I don't get it. What are all those lines? Those points on top of each other are confusing. I don't want to look at this map anymore. It makes me dizzy."

ENVIRONMENTAL SYSTEMS RESEARCH INSTITUTE, INC.

Desktop GIS gives you tools for controlling what appears on a map. For example, there are tools for controlling how big or small features appear and for reducing the number of features, so that just the ones you need are displayed. Other tools let you turn entire map themes on or off, or create more than one view of the information.

How much detail do you need?

The scale of the map determines how big or how small features appear and how much detail you can show. If your audience needs to see a lot of detail in a coastline, the scale must be large enough to show it. A sailor needs to see every cove and harbor to stay on course. A wildlife biologist needs to see all the crenulations in the coastline—its inlets and peninsulas—to identify and preserve good habitat areas. On the other hand, a traveler looking for a good vacation spot along the coast wants to see all the scenic overlooks, access roads, and public beaches, and have a map small enough to unfold in the car. (No map folds *back* quite right.)

One way to control the amount of detail is by *zooming in* and *zooming out*.

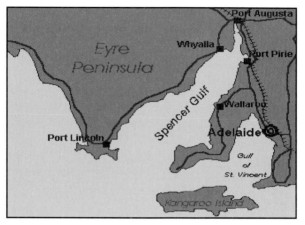

Zooming in enlarges the scale so you can see more detail.

Zooming out decreases the scale and the size of the features. Features like harbors, inlets, and offshore islands become too small to see.

How many features do you need?

No matter how much you zoom in or out, there may still be more features in a theme than you want to show. You don't want to eliminate features, just hide some of them for a while until you need them again. You can do this with a process called *filtering*. For example, you have a theme of cities with populations larger than one million. You want to show only those cities with populations larger than ten million. You simply apply a filter that hides the cities that are too small.

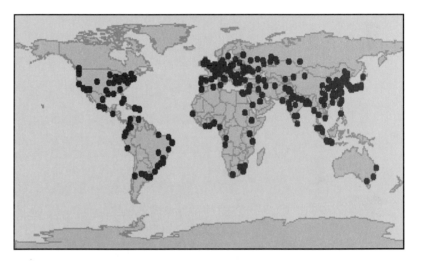

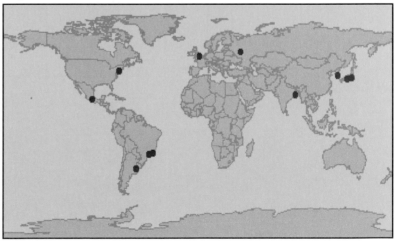

Filtering hides the cities you don't want to show.

Or, suppose you have a theme of properties and you only want to show those that are vacant and zoned for commercial development. Filtering allows you to hide the properties that don't meet your criteria and show the ones that do.

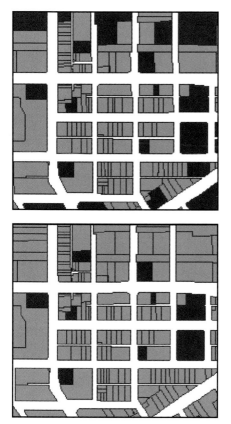

You can display all the vacant properties, or apply a filter to show just those that are zoned for commercial development.

Filtering doesn't remove the features from the map theme. You can display them again, anytime you want, by removing the filter.

How many themes do you need?

GIS databases usually contain many themes of information. You can choose which themes you need to convey your message. To show a citizens' group which streets will flood next time it rains, you might only need to show the streams and the roads they'll overflow onto. However, to present a case for choosing a particular site for a new retail store to potential investors, you'll want to show more themes, including local demographic information, locations of the major roads, the chief competitors, and the best vacant lots.

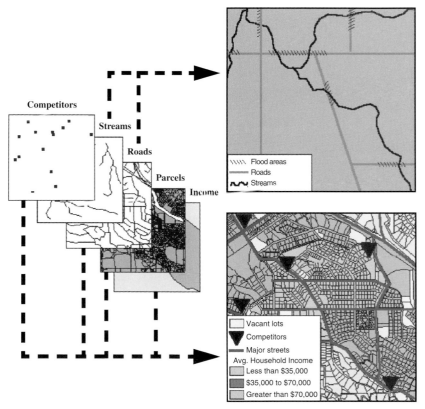

You can use any combination of themes from a GIS database to convey your message.

How many maps do you need?

Sometimes, you really do need to present more information than fits comfortably on a single map. To avoid a cluttered map that's hard to read, you can combine a series of maps in one presentation. This way, you can show separate themes or different portions of the same geographic area, or changes in a geographic area over time.

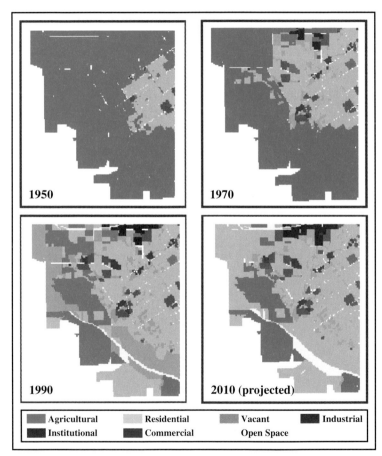

This map series shows past and predicted changes in land use over a sixty-year period.

Organizing and displaying information

OK. Now you've figured out the whos and the whats, but you still have to decide *how* to organize the information on a map and the best way to present it. To present information as accurately as possible, you need to choose the most appropriate symbols. Some colors are commonly accepted as appropriate for certain things, green for plants and blue for water, for instance.

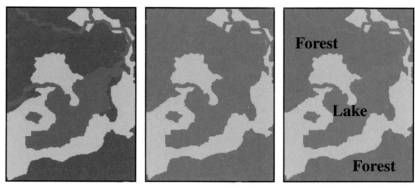

Colors indicate a difference between these area features. Shading the water blue and the forests green conveys more meaning; text labels remove any ambiguity.

On road and topographic maps, a small triangle almost always means a mountain peak. A skull and crossbones likely indicates danger or poison.

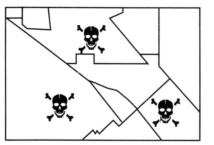 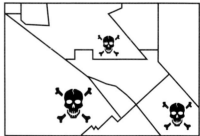

Using a skull and crossbones as symbols to mark toxic spill sites is more meaningful than using plain symbols. Varying the size of the symbol gives more information by indicating that toxic spills come in different sizes.

Line thickness or color can distinguish one type of road from another. A thicker line indicates a wider road; a red line usually indicates a major road.

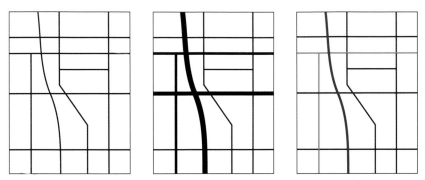

Varying the size or color of these line symbols shows that there is a difference between the roads they represent.

Classifying information

Sometimes you need to present information that has a large number of unique values such as income or population. Since it's too confusing to display each value, you divide them into groups. Each member of a group receives the same symbol. Dividing values into groups, or *classifying,* lets you present a lot of information on a map without overwhelming your audience.

ENVIRONMENTAL SYSTEMS RESEARCH INSTITUTE, INC.

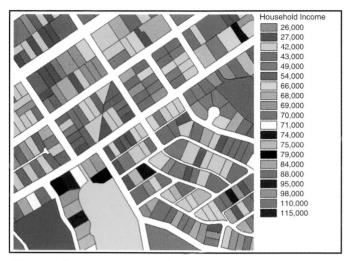

This map shows every unique value for household income with a unique symbol. It's hard for the audience to remember which symbol represents which value and to tell the difference between one symbol and another.

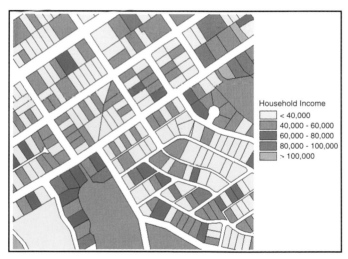

On this map, the household income values are grouped into five classes. Fewer symbols make the information easier to understand.

Some information is more meaningful when you know something about how it ranks, from lowest to highest, such as sales information, housing prices, ages, incomes, temperatures. A *color ramp* uses a range of colors to indicate ranking or order among classes. The colors progress in an orderly fashion from light to dark or from one color to another.

Income classes for households are displayed using a color ramp. The lowest income class (light green) progresses to the highest income class (dark green).

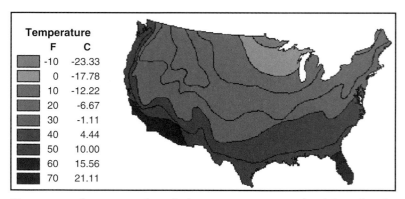

Temperature classes range from the lowest temperatures at the violet end to the highest temperatures at the red end.

Using color ramps to indicate ranked classes makes maps more informative. And the good news: color ramps are easy to create with desktop GIS.

Depending on how you define classes, you can create different maps showing different patterns. For example, if two different methods are used to divide population values into the same number of classes, two different maps would result.

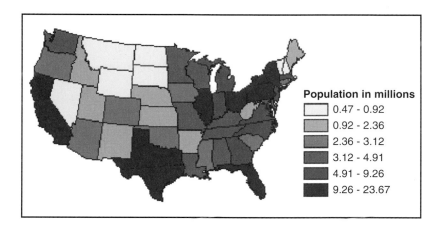

Population in millions
	0.47 - 0.92
	0.92 - 2.36
	2.36 - 3.12
	3.12 - 4.91
	4.91 - 9.26
	9.26 - 23.67

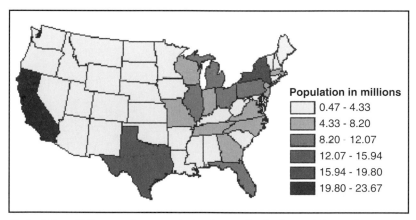

Population in millions
	0.47 - 4.33
	4.33 - 8.20
	8.20 - 12.07
	12.07 - 15.94
	15.94 - 19.80
	19.80 - 23.67

Both of these U.S. maps show population figures for the forty-eight contiguous states grouped into six classes. Each map tells a different story because each uses a different method to classify the information.

On the top map, each class contains the same number of states (eight). On the bottom map, each class contains the same population (one sixth of the total). The patterns formed by the symbols on each map are different, and so is the message conveyed.

When symbols aren't enough

Text, of course, identifies features with more certainty than symbols alone. Green shading on a map may indicate that those features are parks, but which parks? Text tells you the name. Blue may indicate a body of water, but which body? Text tells you the name.

Some text labels do more than name features. They give you the elevations of mountain peaks, the distances between points along a route, and other attribute values.

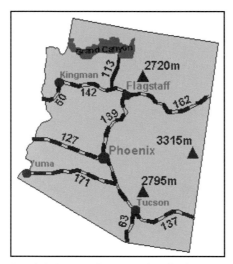

Text labels name features, identify elevations, and indicate distances.

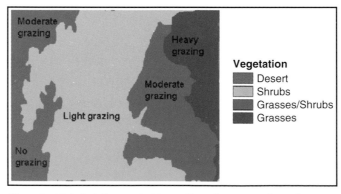

These text labels tell you what type of grazing occurs in each area.

ENVIRONMENTAL SYSTEMS RESEARCH INSTITUTE, INC.

Using charts, tables, and other graphics

Desktop GIS is more than mapping. Using desktop GIS tools, you can create such traditional presentation graphics as pie charts, bar charts, and tables. Not only that, you can include graphics in and add text to the presentation.

Like maps, charts alone are a powerful way of presenting information to others. Used in combination with maps, charts give the audience a different view of the information.

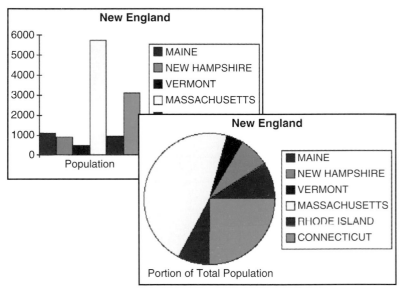

A GIS creates different types of charts as easily as it does maps, from the same data.

Tables can be an effective format for presenting detailed information about a map. You might show selected portions of a table as part of a map presentation or include an entire table in a report.

You may want to enhance your final presentation by adding text and graphics. An arrow here or an explanation there might be just what the audience needs to better understand the message. Placing graphics such as images, scanned pictures, or documents in your presentation adds another dimension.

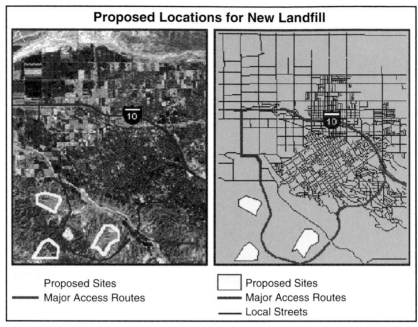

Including an image along with a map of the same area makes a more informative presentation than the map alone.

The dazzling presentation

You know who your audience is, what information to show them, and how to organize and display it. Now you're ready to use desktop GIS to create a presentation that accomplishes your objectives and looks good too.

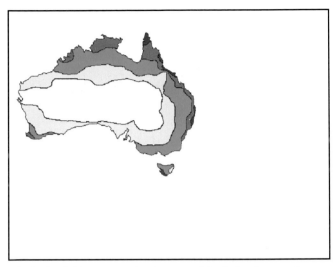

The body of this presentation is a map. It includes the map features you've chosen to show and the symbols used to show them.

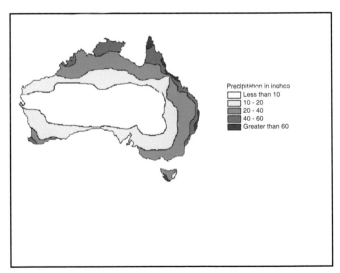

Precipitation in inches
Less than 10
10 - 20
20 - 40
40 - 60
Greater than 60

The legend contains a sample of each symbol and describes what each symbol means.

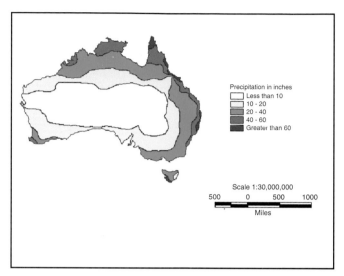

The map scale is traditionally shown in two ways, as a ratio and as a scale bar.

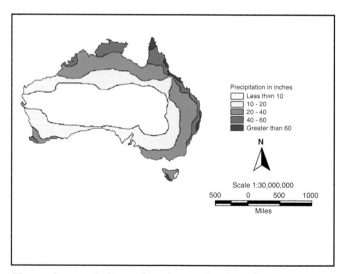

The north arrow indicates how the map is oriented.

ENVIRONMENTAL SYSTEMS RESEARCH INSTITUTE, INC.

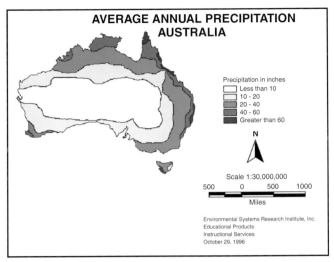

The title identifies the topic of the presentation. Other text can tell who made the presentation and when, as well as where the information was gathered and when.

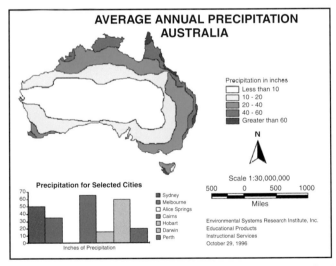

A table, chart, photograph, or document adds information to what's already on the map.

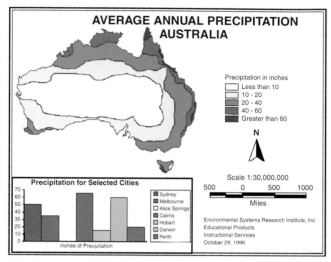

Neatlines provide the finishing touch to a presentation.

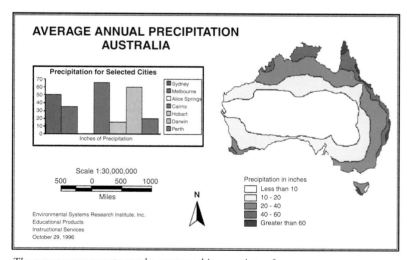

The same components can be arranged in a variety of ways.

The most flexible mapping system

When you create a presentation with desktop GIS, you have the flexibility to change anything, at any stage of the process. You decide what information to present and how much—how much detail, which colors and symbols, and how the final pieces will be arranged. And if your audience or your objective should change, it's easy to make your presentation reflect those changes, without having to start over.

ENVIRONMENTAL SYSTEMS RESEARCH INSTITUTE, INC.

What you need to know about data

A decision is as good as the information

that goes into it.

—John F. Bookout, Jr.

What you need to know about data

Now you know how desktop GIS works. You know that it can help you answer questions and solve problems and present your solutions as dazzling map presentations. All you need now is data.

Deciding what data you need is an important part of any GIS project. Once you know what you need, you need to know where to find it. Once you find it, you need to evaluate it. This chapter gives you information about the types of data you can use with desktop GIS, how to evaluate it, where to get it, and how to make it yourself.

Understanding geographic data

Geographic data is information about the earth's surface and the objects found on it. This information comes in three basic forms: spatial data, tabular data, and image data.

Spatial data—what maps are made of

Spatial data is at the heart of every GIS project, or *application.* Spatial data contains the locations and shapes of map features. Also known as *digital map data,* this is the kind of data you need to make maps and study spatial relationships.

Spatial data includes points that represent such things as shopping centers, banks, and physicians' offices, and lines that represent such things as streets, highways, and rivers. It also includes polygons that represent natural areas and political or administrative areas such as the boundaries of countries, states, cities, census tracts, postal zones, and markets.

Geographic boundaries often come with their areas and perimeters already calculated for you. Street data often includes address ranges along each street.

ENVIRONMENTAL SYSTEMS RESEARCH INSTITUTE, INC.

Tabular data—adding information to geography

Tabular data—the descriptive data that GIS links to map features—is the intelligence behind the map. Tabular data is collected and compiled for specific areas like states, census tracts, cities, and so on, and often comes packaged with feature data.

You probably have some tabular data that's suitable for use with desktop GIS. If you have customer lists or spreadsheets or databases, you can use the GIS to link that information to map features. For example, you can link a sales database with postal code areas so you can map sales volume by postal code. Then if you add commercially available tabular data such as demographic statistics by postal code, you can profile each community.

Some tabular data contains geographic locations, such as addresses, bird nest sites, or places where crimes have occurred. You can use these locations to create map features that can be displayed and analyzed along with other spatial data. You can use the addresses from a customer list to create points on a street map and display them along with the boundaries of your sales territories.

Images—adding another dimension

Image data includes such diverse elements as satellite images, aerial photographs, and scanned data (data that's been converted from printed to digital format).

Images of the earth taken from satellites or airplanes can be displayed as maps along with other spatial data containing map features. You can also use these images as attributes of map features. You could link a satellite image of San Francisco to a map feature so that clicking on the feature would display the image.

Source: EOSAT

A satellite image can be used as a map or as an attribute of a map feature.

Almost any document or photograph can be scanned and stored as an attribute in a GIS database. Photos of houses for sale can be linked to a real estate map; field data forms can be linked to sample sites; and scanned permits can be linked to building sites. Desktop GIS lets you access this information when you need it by simply clicking on a map display.

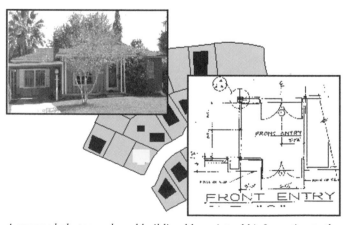

A scanned photograph and building blueprint add information to the map.

Referencing spatial data

The locations of map features are referenced to actual locations of the objects they represent in the real world. The positions of objects on the earth's spherical surface are measured in degrees of latitude and longitude, also known as *geographic coordinates*. On a flat map, the locations of map features are measured in a two-dimensional *planar coordinate system*. Planar coordinates describe the distance from an origin (0,0) along two separate axes, a horizontal *x* axis representing east–west, and a vertical *y* axis representing north–south.

Because the earth is round and maps are flat, getting information from the curved surface to the flat one requires a mathematical formula called a *map projection*. A map projection transforms latitude and longitude locations to x,y coordinates.

Moscow

Moscow

Geographic
Degrees
Latitude: 37° 36' 30"
Longitude: 55° 45' 01"

Universal Transverse Mercator
Meters
X: 412,648.41
Y: 6,179,073.07

Locations are expressed as latitude and longitude on the globe and as x and y coordinates on a map.

This process of *flattening* the earth creates distortions in distance, area, shape, and direction. The result is that all flat maps are distorted to some degree in these spatial properties.

Fortunately, there are many different map projections. They are distinguished by their suitability for representing a particular portion and amount of the earth's surface, and by their ability to preserve distance, area, shape, or direction. Some map projections minimize distortion in one property at the expense of another, while others strive to balance overall distortion.

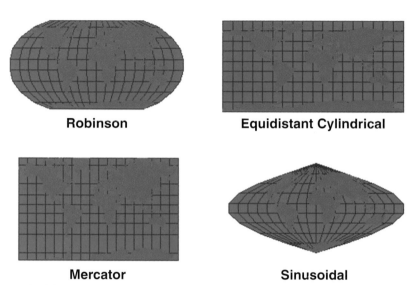

Robinson   **Equidistant Cylindrical**

Mercator   **Sinusoidal**

Each of these world maps uses a different map projection. Different projections cause different distortions.

What to evaluate in data

No matter what the source of the data you use with desktop GIS, you'll need to evaluate the quality and appropriateness of the data for your application. Here are seven issues you need to consider when you select spatial, tabular, or image data for GIS.

Managing distortion

The impact of the distortion caused by a map projection on your work depends on how you will use the data and the size of the area you're covering. If you use measurements to make important decisions, you need to use a projection that doesn't distort whatever you're trying to measure.

If your data covers the entire world, the amount of distortion caused by a map projection may be significant. Knowing the characteristics of the map projection you are using is important, especially when your application involves comparing the shape, area, or distance of map features.

If the data for your area covers a very small part of the earth's surface, such as a small city, then the distortion caused by the map projection you use may be negligible.

Lambert
Conformal
Conic

Area: 8,857,250 m

Albers
Equal-Area
Conic

Area: 8,857,220 m

When a small area is projected, the distortion may be negligible.

Desktop GIS makes it easy to work with spatial data from a variety of sources at the same time. To do so, all the data in your GIS database needs to be in the same map projection so you can use it and display it together. Because of the distortion inherent in *all* map projections, data won't align properly unless it's in the same projection. You might not even be able to view themes together if they're not in the same projection. For example, if your county boundaries are in one map projection and your rivers are in another, the features might appear shifted from their actual locations when you display both themes at the same time.

Spatial data sets in the same projection can be viewed together.

Spatial data sets in different projections do not display together properly.

A desktop GIS can store feature locations as either unprojected geographic coordinates or as projected x,y coordinates. Buying unprojected data gives you several advantages. Since data stored as geographic coordinates can be displayed in any supported projection, you can combine it with data you already have that may be in a different projection. Also, if you use the same data in different applications, you can change projections according to what's appropriate for each application.

Covering your territory

The more specifically you define the geographic area you need to cover, the more precisely you can define how much data you need. If you're deciding where to open a store, you obviously want to know about the tastes and incomes of the people in the immediate area. But if your store will attract customers from a distance away, the area you cover needs to be larger.

If you're studying wildlife habitat, you may need to consider more than one set of boundaries. For example, the flora probably stays put, but the fauna likely ignores those boundaries. Your data may need to cover the territory of wide-ranging and migrating species. If a river flows through your study area, you may need to consider the area where it starts or the portion of it that passes through a large city before it gets to your area.

On the other hand, you don't want geographic information for areas you don't plan to investigate. This information just takes up space on your computer. And the larger the data set, the longer it takes to process.

Getting enough detail

Cities represented as points don't give you information about their shapes. You can't measure their dimensions or find the features, such as roads and parks, contained within them. Rivers represented as lines don't show anything about contours of their banks or changes in width. Buildings as points don't tell you anything about their shapes, nor can you measure the distance between them or between their walls and their lot boundaries.

On the other hand, if you only want to locate your customers, you really don't need to know the shapes of their houses.

Note that the amount of detail with which geographic objects are depicted also influences their attributes. When an entire forest is depicted as a simple area with a boundary, all of the attributes linked to that feature describe the forest as a whole. If a forest is depicted as a set of districts, each one having its own attributes, the attribute information can be more detailed.

When timeliness counts

While physical features like mountains don't change too often, other geographic objects are in constant flux. If you're tracking safe routes through the shifting sand bars of the Mississippi, you want current information. If your application deals with census tracts or postal codes, you want the most up-to-date boundaries. If your application deals with finding customers based on their street addresses, you won't be able to find an address on a new street if the street doesn't exist in the database. The same goes for attributes. If you use demographic data for marketing, you know that good results depend on having the most recent information about potential customers.

When accuracy matters

Some projects require a higher degree of locational accuracy than others. Data may be accurate enough for one use, but not for another. For example, if a line feature representing a road is mapped to within 40 feet of the road's real-world location, it may not be accurate enough for the transportation engineer, but is more than acceptable to the traveler who uses the map to get from one place to another.

Understanding attribute codes

Attribute data is often stored in abbreviated or even cryptic ways. An attribute name might be abbreviated in a table, or six different types of vegetation might be coded as "a" through "f." A catalog that explains the data is called a *data dictionary*. The data dictionary is where you look up the full names of attributes and the meanings of codes. The data dictionary may also include other useful information, like when the data was gathered, the scale of the original source information, the accuracy of locations, and the map projection used. Data that doesn't come with a data dictionary may not be usable.

Data Layer	Name	Source	Date	Projection	Acc	Item	Code	Defin.
Zoning	ZONE	plan.dpt	1994	Mercator	20°	**ZONE2**	AGR	Agricultural
							RES	Residential
							COM	Commercial
							IND	Industrial
							OS	Open Space
							INS	Institutional
Soils	SOILS	SCS	1990	Robinson	200°	**SOIL_TYP**	01	Alo Clay
							02	Delhi Sand

A data dictionary can be a paper or computer document. It contains information about the data.

Compatibility of formats

The format of the data you choose must be compatible with the desktop GIS you plan to use and with any data you already have. The documentation that comes with your desktop GIS software should tell you what spatial, tabular, and image data formats it supports.

Where in the world can you get data?

Data sources abound and data is getting to be less expensive and more available all the time. Your own company or organization may be a source of data. Finding data may be as easy as loading the sample data that comes with your desktop GIS. The best systems include commonly used spatial and tabular data, such as political boundaries and census statistics, as part of their package.

Governments and all kinds of agencies within them collect feature, tabular, and image data. In some countries this data is available to the public at minimal or no cost. In the United States, for example, the Census Bureau's TIGER street-centerline data is one of the foremost spatial data sets for businesses, enabling them to locate their customers on a map.

If you work for local government, you may be able to share data with other departments. If the fire department uses a spatial data set containing streets to determine the best route to take when responding to an alarm, the police department could use the same street data to map crime incidents, and the transportation department, to map automobile accident sites.

Buying data

Vendors collect, package, and sell data for a wide range of applications, from business to natural resources. The spatial data sets they sell contain specific features, such as streets and highways, political and administrative boundaries (states and counties), census areas, postal areas, and marketing areas. The tabular data sets they sell contain specific types of attributes. Business establishment data contains the location, number, type, and characteristics of businesses; census data contains the age, sex, race, income, and housing types of potential customers; health care data contains information about hospitals, physicians, other health care providers, the services they provide, and the demand for those services; and environmental data includes measurements of climate, stream flow, water quality, soil type, and more.

Image data includes over three million images of the earth's surface that have been collected from earth-orbiting satellites. Many of the objects found on the earth's surface can be interpreted and mapped using images.

These data sets are available for areas of all sizes—from a small area defined by a circle around a business location to regions, nations, and the entire world. To help you get started, we've provided a list of data resources, including some data providers, in appendix B.

Creating your own data

Even if you buy data, you may want to make some of your own. With desktop GIS, you can create maps from tabular data that contains locational information. For example, you can map customers from a list of street addresses; map shopping centers from a list containing their x,y coordinate locations; or map wildlife from a list containing the latitude and longitude of each wildlife site.

In addition to creating maps from tabular data, a desktop GIS lets you create maps by drawing shapes over the top of existing maps or images, or create your own attributes by creating new tables or adding columns to existing ones.

Creating features from files

People often have data they don't know is usable for GIS. Businesses have customer files containing information about what customers bought, when they bought it, and what they paid for it. Police departments have accident files containing information about where accidents occurred, what types of accidents they were, and what conditions existed at the time of each accident. Wildlife biologists have files containing information about where nesting sites are located, how many birds are at each site, and the general condition of the birds. Whenever data sets include locational information, you can use them to generate a map.

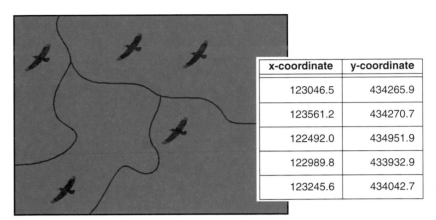

x-coordinate	y-coordinate
123046.5	434265.9
123561.2	434270.7
122492.0	434951.9
122989.8	433932.9
123245.6	434042.7

Desktop GIS can easily map the locations of nesting sites from a file containing their coordinate locations.

Address geocoding

Addresses are actually the most common form of locational information. An address specifies a location in much the same way as a geographic coordinate does. But, addresses are merely text strings containing a house number, street name, and postal code. The GIS needs a mechanism to calculate their geographic location coordinates before you can display them on a map. *Address geocoding* allows you to display tabular data containing addresses as points on a map. To do so, a GIS associates addresses stored in a tabular file with a spatial data set, usually a street network that also contains addresses. The GIS then uses the coordinates of the street features to calculate and assign coordinates to addresses in the file. The result is a map on which each point represents an address location in your file.

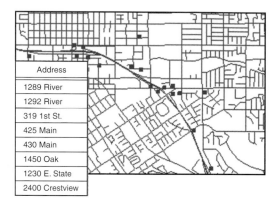

Address
1289 River
1292 River
319 1st St.
425 Main
430 Main
1450 Oak
1230 E. State
2400 Crestview

You can match the restaurant addresses in a file to a street network and show them on a map.

There are countless applications for address geocoding. You can map the addresses of customers, facility sites, club members, retail stores, stops on a delivery route, crime locations, and more. The ability to create map features from files of addresses and other geographic locations is a powerful tool for making better use of the data you already have.

Creating features from shapes you draw

Using desktop GIS tools, you can draw shapes (boxes, circles, points, lines, and polygons) on top of maps and save the shapes as map features. For example, you can draw trade areas on top of a map containing shopping malls. Once you've saved them, you can assign attributes to them and perform desktop GIS operations on them.

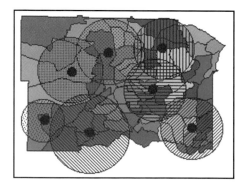

Drawing tools make it easy to define trade areas. Each shape becomes a feature for which you can store such attributes as sales figures and demographics.

ENVIRONMENTAL SYSTEMS RESEARCH INSTITUTE, INC.

Not only can you draw shapes on top of maps, you can also draw shapes on top of images. Images are ideal for creating map features when they have been projected and corrected for photographic distortions. You can trace whatever objects the image contains—rivers, roads, buildings—to create themes of map features. In this way, you can create an entire map database from just one image.

© CNES/SPOT Image

Each object you trace becomes a feature in a map database.

Ready, set, go...   Now you have the information you need to get started with desktop GIS and create your own GIS solutions. In the next chapter you'll see a real application with real data, using a real desktop GIS called ArcView GIS.

Using desktop GIS

Few things are harder to put up with than

the annoyance of a good example.

—Mark Twain

Using desktop GIS

So far, you've seen examples of what desktop GIS can do, how it works, the kinds of questions it can answer, and how you can use it to present information as maps, tables, charts, and more. Now it's time to see a real desktop GIS session with real data being applied to a realistic problem in a realistic situation. The situation may not be your situation, but you will be able to appreciate the process, the steps involved, and the specific GIS tasks performed.

Finding the right place

Finding the best site for expanding a business requires several types of information—information that can be linked with geographic locations. You need to know who is most likely to need your goods or services, and where these potential customers can be found. Once you know where your best customers are, you want to know where to put your business. It needs to be in a place you can afford, where your customers can get to you easily, and far enough away from your competition so your customers don't go there first.

The situation

Wild Outdoors is a business looking to bring its unique goods and services to the Atlanta area. Wild Outdoors specializes in sales and rentals of equipment, such as equipment for skiing, camping, climbing rocks and mountains, and even walking and running. Their customers are out-of-doors enthusiasts who range in age from eighteen to fifty-nine. They have attended some college or hold college degrees. They work mostly in white-collar professions, and live in houses they own or for which they pay higher than average rents. All of Wild Outdoor's best customers have incomes above the national average, and most own a greater than average number of vehicles.

The application

Wild Outdoors executives want to find the best site for a new store in the Atlanta area. To do so, they have identified some important criteria. First, the new site must be located in an area where there are people who match the profile of their current customers. The new site should also be located in a shopping center that's easy to get to. The other shops in the center should complement Wild Outdoors without competing with it.

The executives at Wild Outdoors have unanimously chosen Michael, a member of their marketing team, to perform the site analysis. Michael will use their new desktop GIS software, ArcView GIS, to perform the analysis. ArcView® GIS software has the tools Michael needs, not only to perform the analysis, but also to present his results. And with ArcView's easy-to-use, point-and-click interface, he'll be able to get results in a hurry.

The data

The first step for Michael is to get the right data for his application. He identifies several types of data that he'll need. First, to locate potential customers in the Atlanta area, he chooses consumer data from Equifax National Decision Systems. This data provides information about distinctive consumer lifestyles called *segments*. Each segment is based on an aggregate of consumer characteristics such as income, age, education, and things like the number of vehicles per household. The data contains the number of households belonging to each segment. It's available in a format that is compatible with ArcView and can be licensed for several different-sized geographic areas, such as cities, counties, census blocks, and census tracts. Michael decides to acquire this data for several counties in the Atlanta area at the census tract level.

Michael also chooses shopping center data from National Research Bureau to see where all the shopping centers in the Atlanta area are and what stores are in each center. It comes in a dBASE®-formatted file that ArcView can read and convert to spatial data. The highway data Michael needs to compare locations of shopping centers with major highways comes with the sample data included with ArcView software.

Getting started with ArcView GIS

Loading the data

To begin the session, Michael starts ArcView and creates a new project.

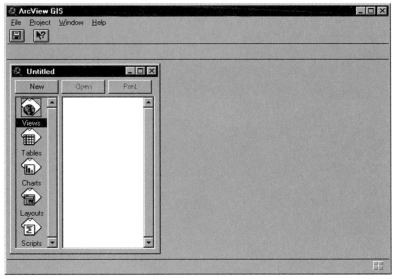

ArcView creates an empty container called a project, *where you save your work during the session.*

Michael loads the consumer lifestyle data into a view. A *view* is a window where you can display themes. In ArcView, a *theme* is a set of map features linked to their attributes. The theme in this case contains the census tracts for Fulton County, the county that Atlanta is in, and an adjacent county, DeKalb.

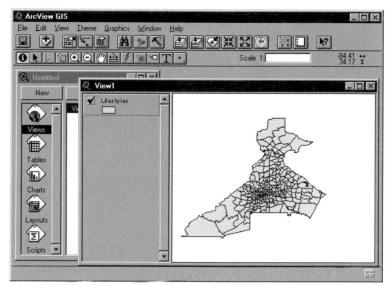

This view *contains one* theme, *called Lifestyles, listed on the left and displayed on the right.*

Displaying the data

When Michael initially draws the theme, the outlines of the census tracts are all displayed with the same symbol. So, he displays the census tracts again based on their county name attribute to differentiate the ones in Fulton County from those in DeKalb.

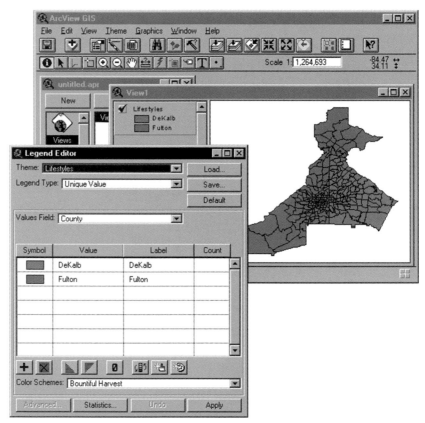

The Legend Editor *lets you choose an attribute and then display the map features based on that attribute.*

ENVIRONMENTAL SYSTEMS RESEARCH INSTITUTE, INC.

Identifying map features

To identify the census tracts and display their attributes, Michael uses ArcView's *Identify* tool. Clicking on a census tract with this tool displays the attributes of that tract in a dialog box.

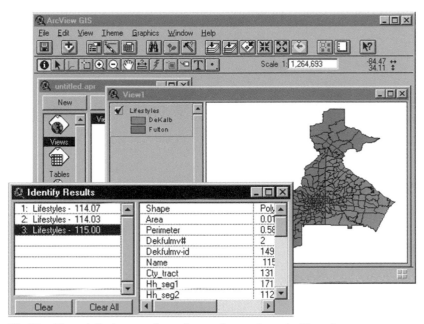

The Identify *tool displays attributes for any feature in an ArcView theme.*

Changing the way information is displayed

As you may recall, the consumer lifestyle data divides households into segments, fifty of them, based on unique consumer characteristics, and stores the number of households per segment for each census tract. By examining the documentation that comes with the lifestyle data, Michael determines that segment number 8, called "Movers and Shakers," most closely matches the profile of the best customers at Wild Outdoors. These customers are over twenty-five years of age. Many are in two-person households with above-average incomes. They hold college degrees and white-collar jobs or are in professional fields.

In order to visualize the distribution of potential customers like these, Michael displays the census tracts based on the segment attribute containing the number of households characterized as "Movers and Shakers."

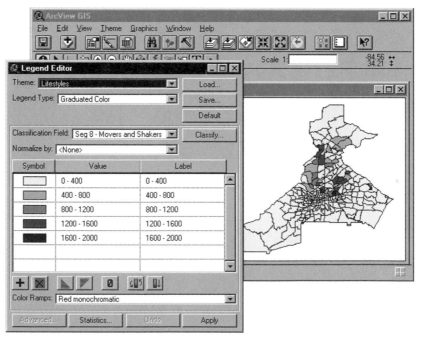

This view shows the distribution of "Movers and Shakers." ArcView lets you classify the values (number of households per census tract) and assign a symbol to each class.

ENVIRONMENTAL SYSTEMS RESEARCH INSTITUTE, INC.

Performing the analysis

The distribution of "Movers and Shakers" gives Michael some idea of the areas where his best potential customers are located. But there are actually three more lifestyle segments that characterize Wild Outdoors customers, segment 4 ("Mid-life Success"), segment 6 ("Good Family Life"), and segment 15 ("Successful Singles"). Michael would like to see where people in all these segments live. Shopping centers located in these areas will be candidates for the new store site.

Finding potential customers

Michael decides that the new Wild Outdoors store must be located where at least 10 percent of the households contain customers from one of these four segments. Based on the average number of households per census tract, Michael has calculated this number to be 500. He uses ArcView's *Query Builder* to find and select the census tracts that meet his criteria.

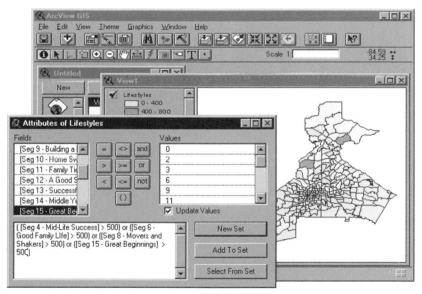

Michael builds a query statement that directs ArcView to search for and select census tracts with at least 500 households in one of four lifestyle segments. Census tracts that meet his criteria are highlighted on the view.

Each census tract shown in yellow has at least 500 households with people from one of the four lifestyles that characterize Wild Outdoors customers. Michael assumes that these people would shop at the new store, were it to be located near them.

At this point, Michael might want to refine his analysis. ArcView has tools he can use to analyze the information in tables, add new information to tables, and then use the new information to display map features. For example, he could rank the four lifestyle segments to reflect the best mix of potential customers and add this information to the database. He could then find all the census tracts that contain 1,000 households of which at least 500 are "Movers and Shakers" and the rest are from the other three segments.

Finding the right shopping center

But, since this is only a preliminary analysis, Michael is ready to add the shopping center data to his view and identify the shopping centers that are located in the selected census tracts. Since the locations and attributes of shopping centers are stored in a dBASE-formatted file, Michael brings this file into ArcView as a table and then creates points for each center based on latitude and longitude information in the table. The shopping centers are added to the view as a theme. Michael draws the shopping center theme to display all the shopping centers on top of the census tracts theme.

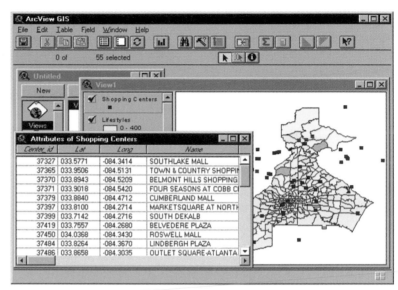

The view now contains two themes: Shopping Centers and Lifestyles.

It's apparent that a number of shopping centers are located in the same areas as potential customers. Michael wants to select only the centers inside the selected census tracts. Although the census tracts and centers are stored in separate themes, ArcView will select features in one theme that are within the selected features of another theme.

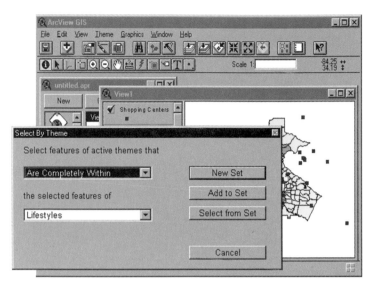

In this case, ArcView searches for and selects the shopping centers that are completely within the selected census tracts.

Selected shopping centers are highlighted on the view. Michael opens the attribute table for the shopping center theme to get information about them.

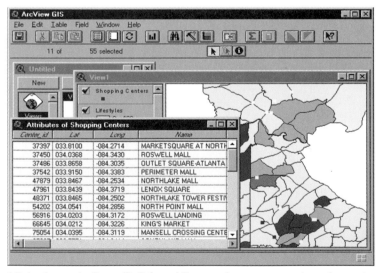

Michael uses the Zoom To Selected Features *button to zoom in to the portion of the view that contains selected shopping centers.*

The eleven selected shopping centers are also highlighted in the table. Michael moves the selected rows to the top of the table with the *Promote* button. Scrolling to the right, Michael can examine all the attributes for these shopping centers, such as gross leasable space, number of stores, type of center, year built, and much more. Michael notices that only some of the centers have available space, so he uses ArcView's *Query Builder* to select only those centers.

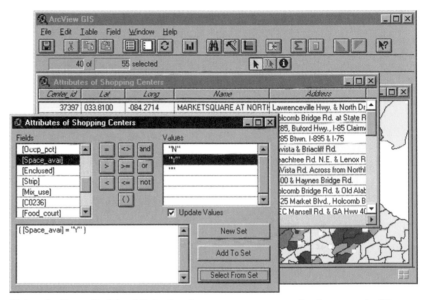

Using the Query Builder, *Michael writes a query statement that instructs ArcView to find and select shopping centers that match his request.*

Only six shopping centers have available space. Since access to the center is another important criterion, Michael adds the highway data to the view so he can determine which centers are located along a major highway.

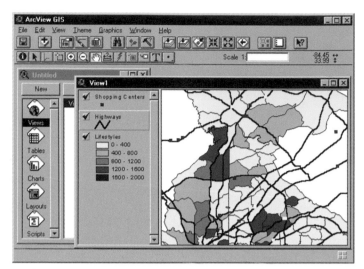

Michael brings the highway data into ArcView as a theme in the view.

Michael can see that each of the six selected shopping centers is easily accessed via a major highway, so none are eliminated at this stage of the analysis.

Evaluating the competition

To further narrow the list, Michael needs to evaluate the competition at each of the six selected shopping centers. Once he determines which centers have the least competition, he can contact a leasing agent to get more information about them, like how much leasable space they have and how much it costs. Then he'll have enough information to make a preliminary presentation to Wild Outdoors' executives.

To evaluate the competition, Michael uses a separate dBASE file that contains the names of anchors (larger stores) and tenants (smaller stores) for each shopping center. Michael brings this file into ArcView as a table. Then he uses ArcView to *link* this table to the Shopping Center Attributes table based on an attribute that both tables contain, the Center_ID attribute. Now when he highlights a center in the Shopping Center Attributes table, all of its anchors and tenants are automatically highlighted in the Tenants table.

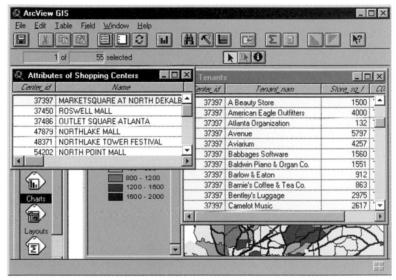

ArcView links tables when there is a one-to-many relationship between them. In this case, one shopping center is linked with many tenants. The two tables are linked based on a column that is common to both, Center_ID.

Michael carefully examines the list of anchors and tenants for each selected shopping center, six in all. As he does so, he notes the number of competitors in each center. Competitors include other outdoor outfitters, sporting goods stores, and stores featuring recreational clothing. Michael would like to compare each center based on the total number of stores and the number of competing stores. So, he uses ArcView to add a column to the Shopping Center Attributes table that contains the number of competitors in each center and uses ArcView's charting capabilities to create a chart comparing the number of stores with the number of competitors.

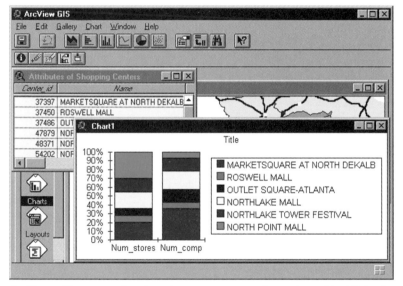

ArcView will create a chart based on one or more columns in a table. Michael identifies two columns for ArcView to chart—the one containing the total number of stores and the one he added containing the total number of competitors.

ENVIRONMENTAL SYSTEMS RESEARCH INSTITUTE, INC.

Not all competitors are created equal

Looking at this chart, Michael can see the balance between the number of stores a center has and the number of competitors it has. But not all competitors are equal. Michael has developed a system to rank competitors, based on how directly they would compete with a Wild Outdoors store. For example, another outdoor outfitter in the same shopping center would provide the most direct competition, while a sporting goods store would provide slightly less direct competition and an outdoor clothing store even less. Michael weights each competitor and calculates a numeric value that represents competition for each of the six shopping centers. The values range from one to nine, where one represents the least amount of direct competition and nine, the most. He adds this information to the Shopping Center Attributes table and then uses it to classify the six shopping centers and display them on a view.

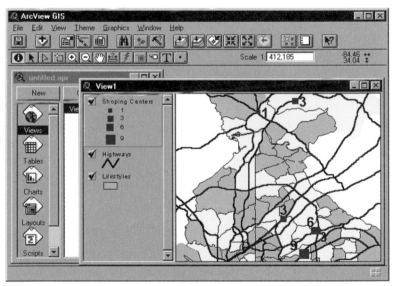

Michael displays the shopping centers along with selected census tracts where potential customers live. The centers are symbolized based on the amount of direct competition they would provide.

Presenting the results

Based on a competition rating of three or less, Michael determines that four out of six shopping centers qualify as potential sites for the new store. However, since one of these centers is adjacent to another one with a higher competition rating, Michael eliminates it, leaving three potential sites. He now has enough information to contact a leasing agent about the size and cost of available spaces and present his preliminary findings to company executives. He would like to show them several pieces of information: a map showing potential sites for the new Wild Outdoors store, along with a chart and a table summarizing the competition at each site. He uses ArcView to create a presentation called a *layout,* which allows him to assemble all the various elements from his analysis to present to company executives. Once the elements are in a layout, Michael can manipulate their size, location, or appearance. Michael can print the layout or use it as an interactive on-screen presentation.

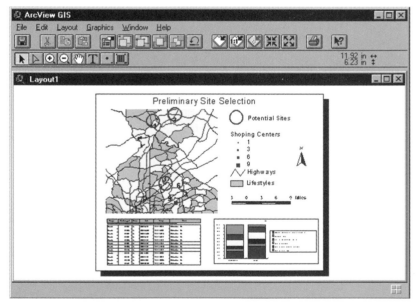

Michael creates a layout that contains a view, a chart, and a table. He adds a number of other elements to aid his presentation, such as a scale bar, a north arrow, and a title.

The outcome

After just one ArcView session, Michael had preliminary results he could present to company executives. The views, tables, charts, and layouts he generated during the session were stored in a single ArcView project that he can reopen and modify any time. Company executives were impressed with the quick results and the opportunity to visualize the data in a variety of ways—as maps, charts, and tables. Although Michael provided each executive with a hard-copy presentation, he also brought his laptop computer with ArcView to the meeting, where executives could make requests and Michael could show them results immediately on the computer screen.

Michael's preliminary analysis gave executives a chance to react and provide additional input. As a result, Michael refined his analysis. The new results gave the company the information it needed to proceed to the next step—an on-the-ground analysis of the best potential sites for a new Wild Outdoors store.

Desktop GIS and you

Desktop GIS. Now you've heard about it. Now you've seen it in action. We hope it will be as valuable to you as it has already been for all sorts of people in all sorts of organizations around the globe.

We've presented only a glimpse of what desktop GIS, and ArcView GIS software, can do. In the remainder of this book you'll see even more applications for desktop GIS, and you'll actually use ArcView to ask questions, get answers, and present your results.

PART 2

Using ArcView GIS

The next 23 chapters teach you how to use ArcView GIS. Each chapter contains exercises that tell a story and guide you through common GIS tasks. You'll learn how to load data into ArcView, then use ArcView's tools to create meaningful displays, answer questions, make measurements, display data at different scales and in different map projections, modify your GIS database, and perform spatial analysis. In addition, you'll learn how to create charts and presentation-quality maps, and how to create some of your own data. You'll see how you can customize ArcView with Avenue, ArcView's own programming language. Finally, you'll look at extensions, programs that add new analysis capabilities to ArcView.

SECTION 1

ArcView GIS basics

The next four chapters introduce you to the basic tasks you perform over and over again in ArcView GIS. Chapter 7 introduces you to ArcView's point-and-click interface. You'll open
an ArcView project and explore its components as well as the
menus, buttons, and tools that let you work with each component. In chapter 8, you'll learn how to load data into ArcView
from a variety of sources (including images) to create a map
display. You'll change the order in which features draw, select
a few features directly from the display, then examine their
attributes. In chapter 9, you'll display features according to
their attributes, change the method used to classify and display
them, and create your own custom classification. Then in chapter 10, you'll learn how to change symbols and label features in
a map display.

SECTION 1:
ArcView GIS basics

Introducing
ArcView GIS

How ArcView GIS is organized

Exploring ArcView GIS

Extensions to ArcView GIS

Getting help

CHAPTER 7

Introducing ArcView GIS

You saw ArcView GIS in action in chapter 6. Now you're ready for a closer look at what it is, how it's organized, and what it does. ArcView is a powerful desktop geographic information system (GIS) made by Environmental Systems Research Institute, Inc. (ESRI). With ArcView, you can load any data that's linked to geographic locations and display it graphically as maps, charts, and tables. Not only that, you can edit the data, change the way it's displayed, append additional data, create some of your own, perform queries to answer specific questions or meet certain criteria, and analyze the information statistically as well as spatially. Then you can show the results as quality graphic presentations to print out or display on your screen.

It's easy to get started with ArcView, and you don't have to be a GIS expert to use it. ArcView's graphical interface lets you point and click to perform almost every operation. ArcView comes with some ready-to-use data so you can get started immediately. In addition, you can get low-cost, off-the-shelf geographic data sets from ESRI and its business partners via the ArcData℠ Publishing Program (see appendix B). If you already have data containing location information, such as addresses, you can load it into ArcView and display it geographically.

How ArcView GIS is organized

ArcView stores the maps, charts, and tables you create in a *project*. A project is a file for organizing all the information you need to do your work. Projects use five types of components (called *documents*) to organize information: *views, tables, charts, layouts,* and *scripts.* Each displays data differently; each has its own related menus, buttons, and tools organized in a unique interface.

Views display sets of geographic data (called *themes*) as interactive maps. Each view has a display area and a Table of Contents that tells you what is being displayed.

Tables display tabular data. Tables containing descriptive information (called *attributes*) about map features are linked to views containing the features they describe. ArcView lets you access the attributes for a feature from a view or from a table.

Charts display tabular data graphically. ArcView charts are fully integrated with ArcView tables and views so you can choose the information to chart by clicking on it in a table or a view.

Layouts are high-quality, full-color presentations that display views, tables, charts, and images as graphic elements on your screen. Layouts can be sent to a printer or plotter to create a hard-copy product. Because ArcView layouts are linked to the data they represent, any changes you make to the data are automatically reflected in the layout, so it's always up-to-date.

Scripts are programs (macros) written in Avenue, ArcView's programming language and development environment. With Avenue™ software you can customize almost every aspect of ArcView, from adding a new button to run a script you write to creating an entire custom application. The version of ArcView that comes with this book does not include scripts, but you'll get a glimpse of what you can do with Avenue scripts in chapter 27.

Extensions are separate software products that run inside ArcView and enhance its capabilities. ArcView comes with extensions that allow you to load CAD drawings, database themes from ESRI's Spatial Database Engine™ (SDE™) database software, ERDAS IMAGINE® images, and JPEG images. Another extension adds digitizer support to ArcView. Other extensions can be purchased independently of ArcView. You can also write your own extensions with Avenue or another programming language. The version of ArcView that comes with this book does not include extensions, but chapters 28 and 29 describe two ArcView extension products, ArcView Network Analyst and ArcView Spatial Analyst, that can be purchased independently of ArcView.

Exploring ArcView GIS

Now you'll take a closer look at ArcView by opening an ArcView project and exploring each of its components. The data you'll work with in this project is World data that comes with ArcView. If you haven't yet installed the CD–ROM that comes with this book, see appendix D.

Exercise 7a

1.   If you are running Microsoft® Windows® 95 or Windows NT™ version 4.0 or greater, hold down the Start button. From the Programs menu, go to the Getting to Know ArcView GIS menu and select "Getting to Know ArcView GIS."

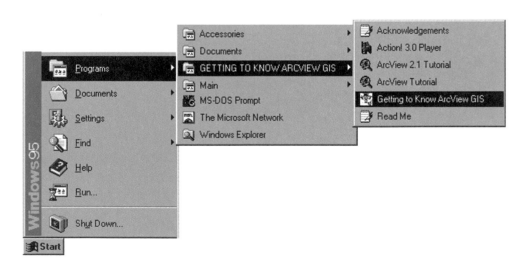

If your machine is running Windows version 3.1 or a Windows NT version lower than version 4.0, double-click on the Getting to Know ArcView GIS program group to open it. Then double-click on the Getting to Know ArcView GIS icon.

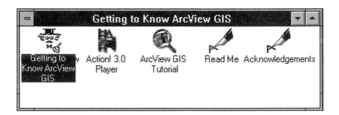

The Getting to Know ArcView GIS window displays.

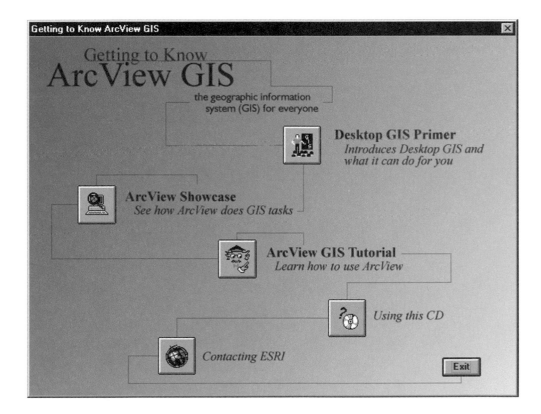

Now you'll start ArcView.

2.   Click the ArcView GIS Tutorial button. The Getting to Know ArcView GIS window disappears. It may take a few moments for ArcView to start, depending on the speed of your computer. You see two windows, a large one entitled ArcView GIS and a small one, the Project window. The title bar for this window displays the name of the current project, in this case, "Untitled." This project is empty because there is no data associated with it yet.

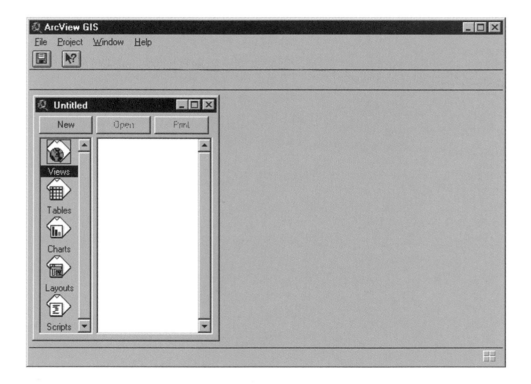

At the top of the ArcView GIS window is a menu bar with four pulldown menus: File, Project, Window, and Help. These menus are available when the Project window is active.

Working with windows in ArcView. Each ArcView project, view, table, chart, layout, and script is contained in its own window. You can have any number of windows open at the same time, but only one window can be active. You make a window active by clicking inside it, clicking on its title bar, or choosing it from the Window pulldown menu. You can resize and move windows using the mouse, using the window icons (to the left and right of the title bar), or using the choices in the Window pulldown menu. For more information, search for this Help Topic: *ArcView's user interface.*

Below the menu bar is the button bar. It currently contains two buttons, Save Project and Help. Buttons offer you quick access to ArcView functions. To find out what a button does, place the cursor over the button. ArcView describes its function in the status bar at the bottom of the ArcView window. (If you are running Windows 95 or Windows NT, you may also see a yellow box with the name of the button appear directly below the button.)

Below the button bar is the tool bar. It doesn't contain any tools yet.

To open an existing project, you'll use the File menu.

3. Click on the File menu to display its list of choices. One of the choices is Open Exercise. This choice doesn't exist in the standard ArcView interface; it's been added for use with this book. You'll use it to open any of the exercise projects included in this book.

4. Click on the Open Exercise choice. You see a scrolling list entitled "Exercises." This list contains all the exercises in this book. Each exercise is linked to an ArcView project. Clicking on an exercise in the list opens the project it's linked to.

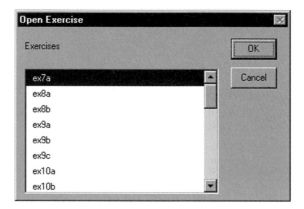

5. In the Exercises list, click on "ex7a" to open a project named *ex7a.apr.* (All ArcView projects have a *.apr* ending, which stands for "ArcView Project.") When the project opens, you see that the name *ex7a.apr* replaces the name *Untitled.*

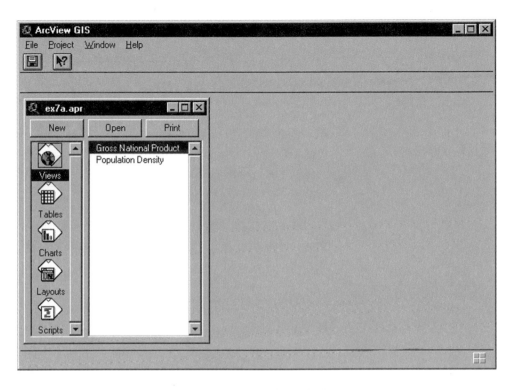

Now take a closer look at the Project window.

The Project window has icons along its left side, Views, Tables, Charts, Layouts, and Scripts, representing each document type. (You won't see the Scripts icon shown here because the version of ArcView you're using has been modified for this book. The Scripts icon is available when you load a standard version of ArcView.)

Currently, the Views icon is highlighted. The buttons at the top of the window, New, Open, and Print, let you create new views, open existing views, and print a view that's highlighted in the list. These buttons change for each document type.

On the right side of the window, you see the names of the two views currently contained in this project. The Gross National Product view is highlighted.

Next you'll highlight and open both views.

6.   Hold down the Shift key and click on the Population Density view. Now both views in the list are highlighted. Click the Open button at the top of the Project window to open them. Two view windows open, and ArcView's interface (menus, buttons, and tools) changes to reflect the *view* document type. You can tell the Population Density view is active because its title bar is highlighted and its window is in the foreground.

                                              ENVIRONMENTAL SYSTEMS RESEARCH INSTITUTE, INC.

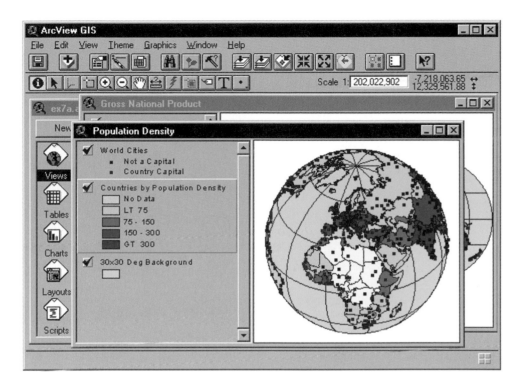

The Population Density view displays three sets of features, called *themes*. Each theme has a title and a legend that appear in the view's Table of Contents, on the left side of the view window. The World Cities theme represents cities of the world as point features. The Countries by Population Density theme shows countries (polygon features) divided into five groups, or classes, based on their population values. Each class is displayed using a different symbol. (Notice that some of the countries display in yellow. This indicates that they've been selected as a separate group. By default, ArcView highlights selected features in yellow.)

In the background is a theme showing a grid of latitude and longitude.

7. Click on the title bar of the Gross National Product view. It becomes the active view.

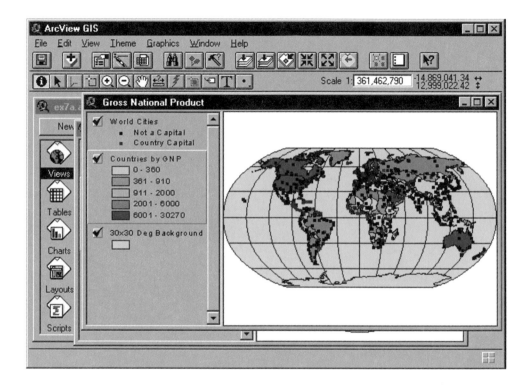

Again you see a theme of world cities, a theme of countries, and a theme of latitude and longitude; however, in this view, the countries are shown according to their gross national product (GNP).

The shape of the world is different in the two views because each view is displayed using a different map projection. In ArcView, you don't have to set a map projection to work with data in a view, although you can choose from a number of different map projections.

You'll close the Gross National Product view and leave the Population Density view open.

8. Click on the icon in the upper left corner of the Gross National Product window and choose Close from the menu. The view window closes.

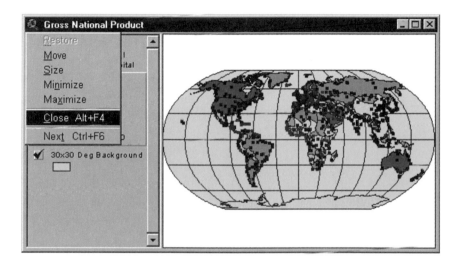

Now you'll look at the attributes that are linked to the features in the Countries by Population Density theme. ArcView stores these attributes in a *theme attribute table* or, simply, *theme table*.

Notice in the Table of Contents that the area containing the name and legend of the Countries by Population Density theme appears raised, indicating that this theme is active. Many of the operations you perform in a view work only on active themes.

Now you'll open the theme table for the active theme.

Opening a theme table. Each theme of geographic data in a view has a table that stores attributes describing the features it contains. To open a theme's attribute table, choose Table from the Theme menu or click the Open Theme Table button on the View button bar. Once you open a theme's attribute table, it appears in the list of tables in the Project window, where it can then be opened using the Open button in the Project window. For more information, see chapter 8 or search for these Help Topics: *Working with tables in a project, Open Theme Table.*

9. Click the Open Theme Table button on the View button bar. The Countries by Population Density table opens.

The table is now the active document, and ArcView's interface changes to display the menus, buttons, and tools you use for working with tables.

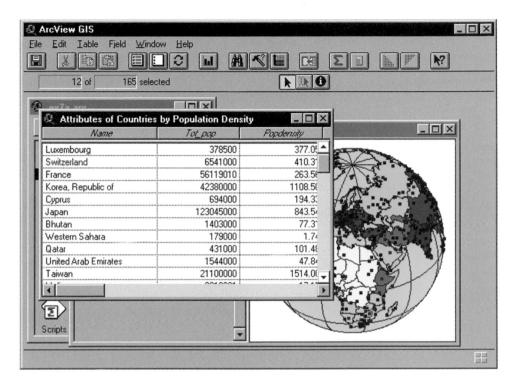

The table contains information about the countries of the world. The information in the column entitled "Popdensity" was used to divide the countries into the five classes shown in the view.

Recall that some of the African countries were highlighted in yellow in the view. The attributes corresponding to those countries are also highlighted in the table, but you can't see them unless you scroll down in the table or use the Promote function to bring them to the top of the table.

10. With the Table window active, click the Promote button. (Or, click on the Table menu to display its choices, then choose Promote.) ArcView moves the highlighted attributes to the top of the table.

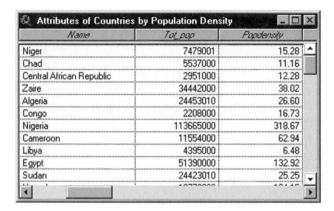

Name	Tot_pop	Popdensity
Niger	7479001	15.28
Chad	5537000	11.16
Central African Republic	2951000	12.28
Zaire	34442000	38.02
Algeria	24453010	26.60
Congo	2208000	16.73
Nigeria	113665000	318.67
Cameroon	11554000	62.94
Libya	4395000	6.48
Egypt	51390000	132.92
Sudan	24423010	25.25

11. Scroll to the right. Notice that the table contains attribute information on female and male life expectancy for each country. Next you'll open a chart showing the life expectancies for the highlighted African countries.

12. Click on the title bar of the Project window to make it active, then click on the Charts icon. The Average Life Expectancy chart is highlighted in the list. Click the Open button. The Average Life Expectancy chart opens.

The chart is now the active document, and ArcView's interface changes to display the menus, buttons, and tools you use for working with charts.

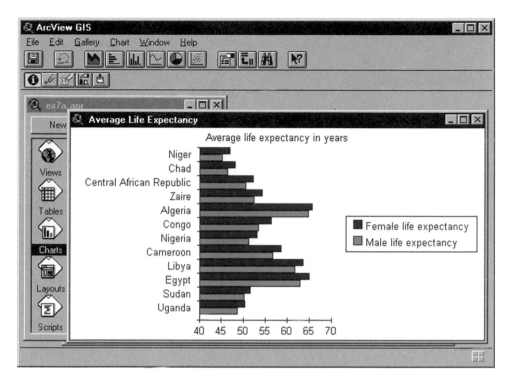

Charts represent tabular data graphically. This chart was created using the values in the Female life expectancy and Male life expectancy columns (fields) in the Countries by Population Density table. Only the countries highlighted in the view are represented in the chart. You'll learn how to create, modify, and query charts in chapter 21.

Next you'll open an ArcView layout.

13. Click on the Project window's title bar to make it active, then click on the Layouts icon. The Population Demographics layout is highlighted in the list. Click the Open button to display it.

The layout is now the active document, and ArcView's interface changes to display the menus, buttons, and tools you use for working with layouts.

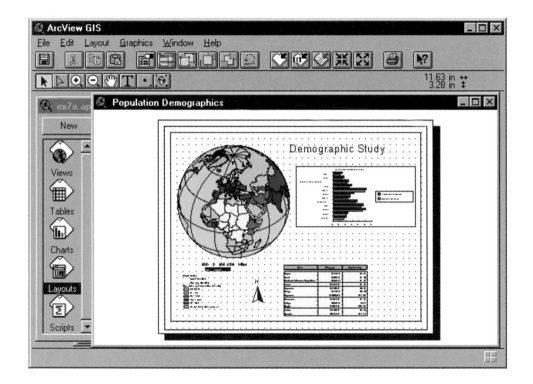

This layout contains the Population Density view and its legend, the Countries by Population Density table, the Average Life Expectancy chart, and additional graphics and text. You can think of an ArcView layout as a canvas or page where you place project components and other graphics to create a presentation-quality display or paper map. You'll learn how to create layouts in chapter 22.

Next you'll close all the windows.

14. From the File menu, choose Close All. Only the Project window remains open and active.

If you want to go on to the next chapter, leave ArcView running. Otherwise, choose Exit from the File menu.

As with any software package, it's important that you save your work regularly. With the version of ArcView that comes with this book, you won't be able to execute a save operation. But you should know how to save, and what happens if you forget to save.

Saving your work in ArcView. In ArcView, you can save the work you do on a view, table, chart, or layout by saving the project that contains it. To save the project, choose Save Project from the File menu, click the Save Project button on the button bar, or press CTRL+S. To save your work to a project with a different name, make the Project window active, then choose Save Project As from the File menu. For more information, search for these Help Topics: *Saving your work, Save Project, Save Project As*.

You've seen an ArcView project and some of the documents it can contain (views, tables, charts, layouts). You've also seen that each document type has its own interface containing menus, buttons, and tools.

In the chapters that follow, you'll perform specific GIS tasks using views, tables, charts, and layouts in a project. After you've worked all the exercises, you'll be ready to tackle your own ArcView project.

Extensions to ArcView GIS

As you become more familiar with ArcView, you may want to automate certain processes, add new functions, or customize the interface to suit your applications. You can use Avenue, ArcView's scripting language, to do your own programming, or you can load ArcView extensions, which are optional software products that are loaded into ArcView while it's running to give it extended capabilities.

To load an extension, you select Extensions from the File menu (it's grayed-out in this version of ArcView). A dialog box appears, listing the extensions available on your computer. You check the boxes next to the ones you want to load.

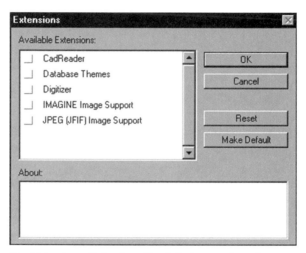

The Extensions dialog box. The default extensions that come with ArcView are shown.

The CADReader extension allows you to load CAD drawings from many standard CAD software packages. The Database Themes extension gives ArcView the ability to request themes from ESRI's SDE (Spatial Database Engine) product. The Digitizer extension allows you to use a digitizer as well as a mouse in ArcView. The IMAGINE Image Support and JPEG (JFIF) Image Support extensions give ArcView the ability to load two popular image types.

ENVIRONMENTAL SYSTEMS RESEARCH INSTITUTE, INC.

In addition to the default extensions, other extensions, such as the ones you'll see in section 9, can be purchased separately. You can also create your own extensions using Avenue and other programming languages.

Getting help

With ArcView's extensive online help system, you can get answers to most of your questions about ArcView. ArcView offers several kinds of help. For example, you can get context-sensitive help by clicking on the Help button and clicking on a menu choice, button, tool, or document. A Help window gives a description of the item you clicked on.

By selecting Help Topics from the Help menu, you can choose from the Contents, Index, or Find tabs. Use the Contents tab to view the Help contents arranged in logical order like a book. Use the Index tab to select a topic from an alphabetical list. Use the Find tab to bring up topics that contain a specific word or words.

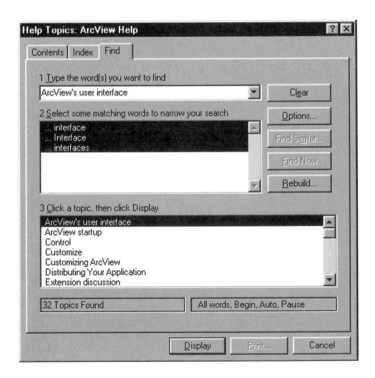

The teal-colored boxes throughout this book contain references to Help Topics. To access these topics, click on the Find tab and type the topic name in the input box, then select the topic from the scrolling list at the bottom.

Getting Help on Help. To learn how to use ArcView's online help system, click on the Help menu (available on every ArcView interface), then choose How to Get Help to display the ArcView Help Contents menu.

ENVIRONMENTAL SYSTEMS RESEARCH INSTITUTE, INC.

SECTION 1:
ArcView GIS basics

Getting data into ArcView GIS

Understanding data sources

Adding themes to a view

Understanding theme tables

Getting data into ArcView GIS

ArcView GIS links sets of features to their attributes in *themes* and manages them in a *view* (chapter 7). You can create an ArcView theme from a variety of geographic data sources, such as spatial data, computer aided design (CAD) drawings, images, and tabular data. In this chapter, you'll create themes from several types of spatial and image data sources.

Understanding data sources

Geographic data sources fall into two categories: feature data sources and image data sources. As you might have guessed from the names, feature data sources contain features, whereas image data sources (which are basically pictures) do not. ArcView supports these data sources for feature-based themes: ARC/INFO® coverages, SDE database layers, ArcView shapefiles, and CAD drawings.

Feature data sources

ARC/INFO is GIS software from ESRI, the creators of ArcView. ARC/INFO stores sets of features and their attributes in its own format, called a *coverage*. A coverage can be represented as a theme in ArcView. Some ARC/INFO coverages contain more than one type of feature; ArcView requires a separate theme for each type.

SDE *layers* are similar to ARC/INFO coverages, but they are stored on a machine running ESRI's Spatial Database Engine software, which acts as a database server. ArcView can request SDE layers from the database server and display them as themes if the ArcView Database Themes extension is loaded.

Shapefiles are ArcView's own format for storing features and attributes. You create shapefiles by converting other spatial data sources (such as ARC/INFO coverages), by drawing shapes in themes you create, or by using tabular data containing location information (see chapters 23

ENVIRONMENTAL SYSTEMS RESEARCH INSTITUTE, INC.

and 25). There are a couple of advantages to using shapefiles: they display more rapidly in a view than other spatial data formats, and you can edit a theme that's based on a shapefile. Suppose you want to edit a theme that's based on an ARC/INFO coverage. No problem. Just convert it to an ArcView shapefile.

CAD drawings are another type of spatial data you can use to create themes. A CAD drawing typically contains many entities (feature types) in a single layer. As with ARC/INFO coverages, ArcView requires a separate theme for each entity. To edit a theme based on a CAD drawing, you have to convert it to a shapefile. (Before you create a theme based on a CAD drawing, you must install the optional CADReader extension.)

Image data sources

Image data sources include satellite data, scanned data, photographs, and ARC/INFO grids. (An ARC/INFO grid is a spatial data format that represents features as a group of cells in a regular grid or matrix.) ArcView supports images and grids for display only. However, if you have the ArcView Spatial Analyst extension loaded, you can also create and analyze grid themes in ArcView (see chapter 29).

Adding themes to a view

Suppose that you work for the City Maintenance Department, which plans to add some utilities and upgrade others in a recently renovated part of town. You've been asked to create a map showing the existing utilities to use for planning the additions and upgrades. There is currently no single map that shows all the utilities. Your task is to locate the necessary data sources and add them to a view as themes so you can display them together.

Exercise 8a

1. If necessary, start ArcView. From the File menu, choose Open
    Exercise. In the Exercises scrolling list, select "ex8a," then click OK.
    Because no views have been created yet, you see an empty Project
    window.

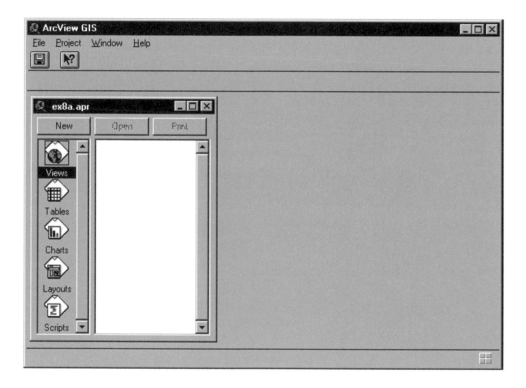

2. With the Views icon highlighted, click the New button. A new,
    empty view window, View1, opens. (In chapter 11, you'll learn how
    to rename the view window.) Drag the View1 window to the center
    of the ArcView window. You can resize and reposition this window
    any time you need to.

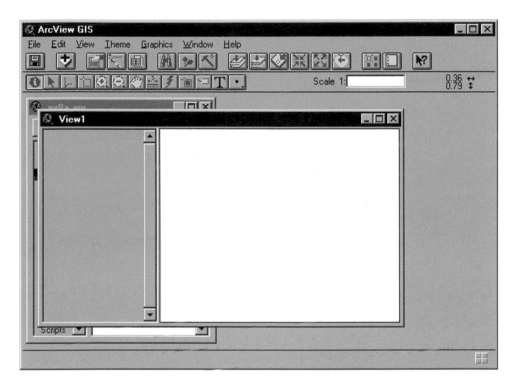

The gray area on the left of View1 is the Table of Contents. It's empty now, but when you add a theme to the view, its name, the symbol used to draw it, and a check box indicating whether it's currently displayed will appear there.

3. From the View menu, select Add Theme. The Add Theme dialog box displays.

4. From the Drives list (bottom right), select the drive where you installed the data for this book, then navigate to *\gtkav\data\ch08* in the Directories list. ArcView lists the geographic data sources available in this directory. When "Feature Data Source" is selected in the lower left drop-down list (Data Source Types), only data sources containing features (e.g., points, lines, polygons) are listed.

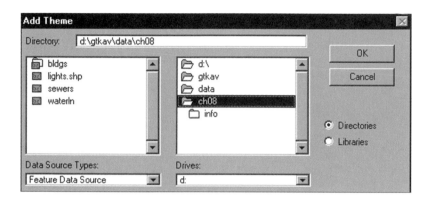

You see four data sources: bldgs, lights.shp, sewers, and waterln. The first of these, bldgs, is an ARC/INFO coverage containing more than one type of feature. It appears with a folder icon in the list. Later, you'll open the folder to see the feature types. The second, lights.shp, is an ArcView shapefile (*.shp* is the default file extension given to ArcView shapefiles). The other two data sources are ARC/INFO coverages.

Adding a theme from a feature data source. The Add Theme dialog box lets you add a theme from a feature data source. Double-clicking on the name of the data source adds it to the current view as a theme. By holding down Shift, you can select and add more than one data source at once. When a feature data source, such as an ARC/INFO coverage, contains more than one feature type, it appears with a folder icon in the Add Theme dialog box. Single-clicking on the icon lists all the feature types for that data source. You must create a separate theme for each feature type. For more information, search for these Help Topics: *Adding a theme to a view, Add Theme.*

You'll add a theme from the waterln data source.

5. Double-click on "waterln" to add it to the view as a theme. The theme's name and a symbol appear in the Table of Contents.

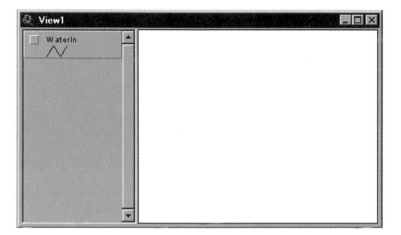

Your view now contains one theme, Waterln. By default, ArcView doesn't draw the theme. To display the theme, you'll turn it on by clicking on its check box.

Turning a theme on just allows it to display. A theme doesn't have to be turned on for you to perform ArcView operations on it. Turning a theme off doesn't remove it from the vlew.

---

6. Click on the check box next to the Waterln theme name. ArcView draws the features in the theme (lines) using the current symbol.

When you add a theme to a view, ArcView randomly assigns a color to the theme. Therefore, the Waterln theme may be a different color in your view. You'll learn how to choose colors in chapter 10.

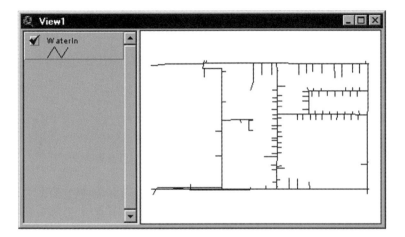

Next you'll add themes based on the bldgs, lights.shp, and sewers data sources.

7. Click the Add Theme button to display the Add Theme dialog box. If necessary, select the drive where you installed the data for this book, then navigate to *\gtkav\data\ch08* in the Directories list. You see the same list of data sources.

8. Click once on the "bldgs" folder icon to open it.

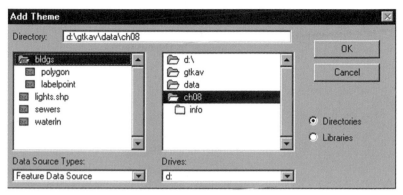

You see two feature types listed, polygon and labelpoint. (A "labelpoint" in ARC/INFO is a point used to identify each polygon.) You want to create a polygon theme to represent buildings, so you'll choose the polygon data source.

9. Click once on "polygon" to highlight it. Then hold down the Shift key and click once on "lights.shp" and once on "sewers." All three data sources are highlighted.

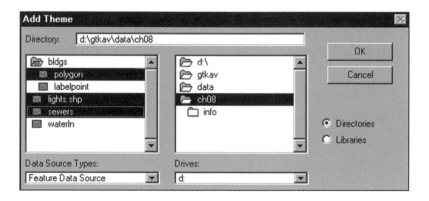

10. Click OK to add these three themes to the view, then click on each check box to draw each theme.

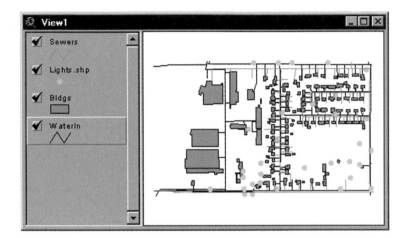

Your view now contains three additional themes, the Sewers theme containing line features, the Lights.shp theme containing points, and the Bldgs theme containing polygons.

By default, the name of the theme is the same as the name of the data source used to create the theme. Suppose you want to change the name of the theme to something more significant. You can do this in the Theme Properties dialog box.

11. With the WaterIn theme active, select Properties from the Theme menu. The Theme Properties dialog box displays.

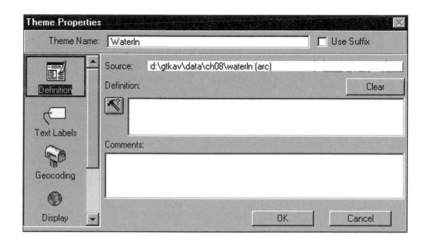

12. In the Theme Name text box, change the name of the theme to "Water Lines" and click OK. The name of the theme is changed in the Table of Contents.

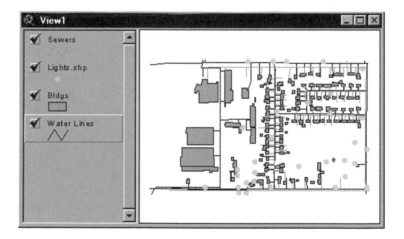

Now you can use the themes to plan for upgrading utilities. But first, you'll add an aerial photograph of this part of the city to serve as a backdrop.

13. Click the Add Theme button. If necessary, select the drive where you installed the data for this book, then navigate to *\gtkav\data\ch08* in the Directories list. Click on the down arrow for the Data Source Types list, then select "Image Data Source." The aerial photograph image source appears in the list on the left side of the dialog box. (The *.bil* ending indicates a type of image format.)

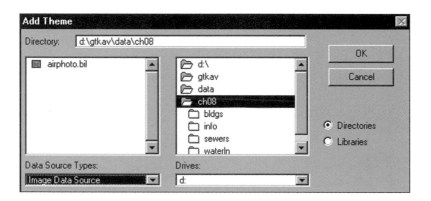

14. Double-click on "airphoto.bil." ArcView adds the aerial photograph image to the view. Click on the check box for the Airphoto.bil theme to turn it on. ArcView draws the photograph as a black-and-white image in the view.

The image draws on top of the other themes. That's because ArcView first draws the theme listed at the bottom of the Table of Contents, then draws each theme listed above it. Thus, the Airphoto.bil theme draws last. You can change the drawing order by dragging themes up or down in the Table of Contents.

You want the image to display in the background (behind the other themes) so you'll drag it to the bottom of the Table of Contents. To do so, you must first make the Airphoto.bil theme active.

Understanding active themes. An *active* theme appears raised in the Table of Contents. Many operations you perform in a view work only on active themes. You make a theme active by clicking on its name or legend symbol in the Table of Contents. To make more than one theme active, hold down the Shift key, then click on the name or legend symbol of each theme you want to make active. For more information, search for this Help Topic: *Making a theme active.*

Notice that the Water Lines theme is currently the active theme.

15. Click once on the Airphoto.bil theme in the Table of Contents to make it active. Now it appears raised in the Table of Contents.

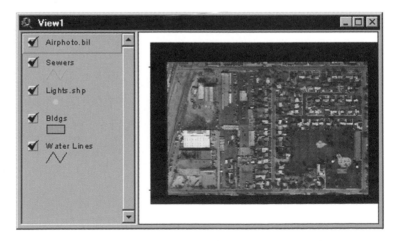

Next you'll drag it to the bottom of the Table of Contents.

16. Place the cursor on the Airphoto.bil theme's name (or the raised gray area surrounding it), hold down the mouse button, move the cursor to the bottom of the Table of Contents, then release the button. ArcView draws the image theme first this time, then draws all the other themes on top of it.

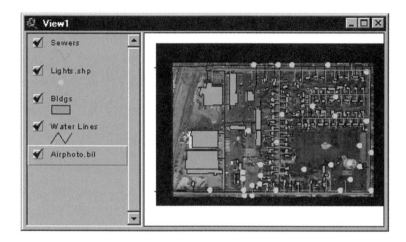

Copying themes between views. There may be times when you want to use the same theme in several views. You could re-create the theme from the data source, or you could copy the theme from one view to another. To copy a theme, you make the theme active and select Copy from the Edit menu. Then you open the view where the theme will be copied and select Paste from the Edit menu. You can copy themes to the same view, a different view in the same project, or a different view in a different project. For more information, look for this Help Topic: *Cutting, copying and pasting themes.*

You can see how easy it is to create a view and add themes to it from a variety of data sources. Once you've added themes to a view, you can change the appearance of the view by turning themes on or off and by moving themes up or down in the Table of Contents. You can also change the name of a theme in the Theme Properties dialog box.

If you want to go on to the next exercise, leave the project open.

Understanding theme tables

When you add a theme based on a feature data source, a *theme attribute table* or, simply, *theme table,* is also added to the project. A theme table contains descriptive information about the features in the theme. The theme table is formatted in rows and columns, called *records* and *fields,* respectively. Each record represents a single feature in the theme; each field contains all the values for an attribute. Because records are linked to the features they describe, you can access them by clicking on a feature in the view, or you can find a feature in the view by clicking on its record in the table. (To review the relationship between features and attributes, see chapter 2.)

The City Maintenance Department has decided to dig trenches for sewer lines on some of the properties. Your task is to retrieve the address information for these properties so notification letters can be sent to their owners. The Bldgs theme attribute table contains the address information you need. You'll make this theme active, then open its attribute table.

Exercise 8b

1. If *ex8a.apr* is open, continue. Otherwise, choose Open Exercise from the File menu. In the Exercises scrolling list, select "ex8b," then click OK. When the project opens, you see a view with four feature-based themes, and an image theme in the background.

2. Click on the Bldgs name or its legend symbol in the Table of Contents to make it active. The theme appears highlighted in the Table of Contents.

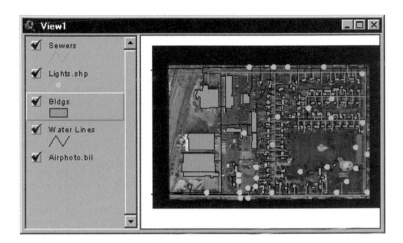

3. Click the Open Theme Table button on the View button bar. A table window opens containing the attributes of the Bldgs theme. When the table opens, you see the first five fields, Shape, Area, Perimeter, Bldgs#, and Bldgs-id. The Shape field tells you the type of feature (e.g., point, line, polygon) the theme represents.

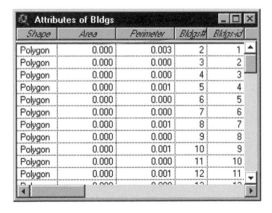

The table also contains addresses and owners for all the buildings in the theme. To see them, you'll use the scrolling bars.

4. Using the scroll bar at the bottom of the table, scroll to the right. The address information is stored in the Address, City, State, and Zip fields; the owner names are stored in the Owner field. (Later you'll resize the table so you can see these fields at the same time.)

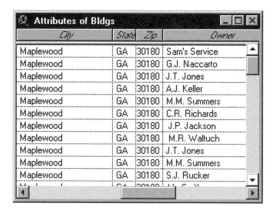

You know that the city plans to dig trenches for sewer lines on the properties of large buildings along the left side of the view. You'll select these buildings by clicking on them in the view.

Before you select the buildings, you'll resize and reposition the view and table so you can see both of them at the same time.

5. Make View1 active by clicking on its title bar. Move it to the upper left corner of the ArcView window, then resize it so it fills the upper portion of this window.

6. Make the Attributes of Bldgs table active. Move it to the lower left corner of the ArcView window, then resize it so it fills the lower portion of this window.

You'll change the table display to show the address and owner information.

7. Using the scroll bar at the bottom of the table, scroll to the right until you see the Address and Owner fields. Your windows should look like this:

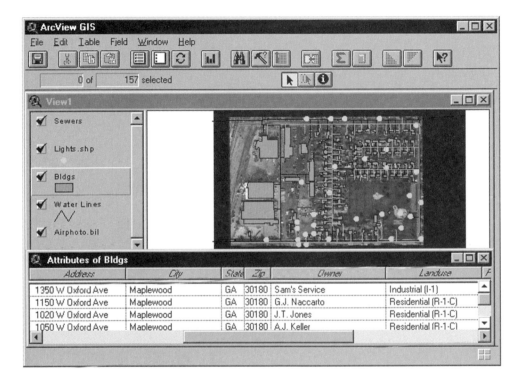

Now you'll use the Select Feature tool to select the large buildings along the left side of the view.

8. Make the view active by clicking on its title bar. In the View tool bar, click on the Select Feature tool, then click on the large building in the upper left corner of the view. The building highlights in the view and the corresponding record highlights in the table. ArcView scrolls the table so the highlighted record displays at the top of the table.

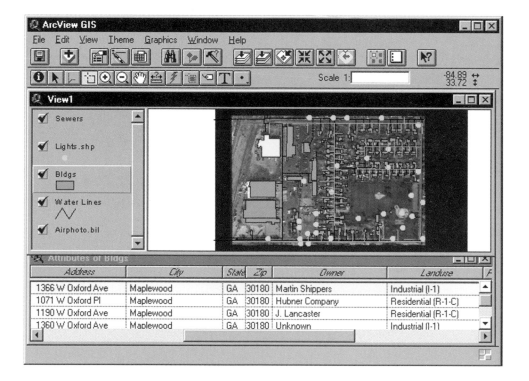

The first building is selected. Now you want to highlight the rest of the large buildings along the left side of the view.

9. Hold down the Shift key, then click on each of the other large buildings along the left side of the view. ArcView selects and highlights the buildings (there are four in all) in the view and their corresponding records in the table.

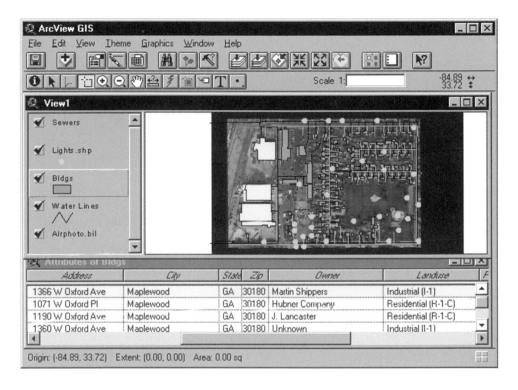

Because the table is large, you can't see all of the highlighted records. To see the highlighted records together in the table, you'll promote them.

10. Make the table window active by clicking on its title bar. Then click the Promote button on the Table button bar. The highlighted records display at the top of the table.

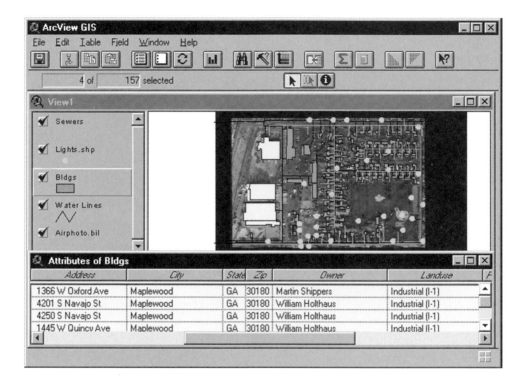

The highlighted records contain address information for the selected buildings. This information can be used to notify owners about the city's plans to put sewer lines in place on these properties.

Selecting features in a view allows you to access their attributes in the theme table. You'll learn other ways to select features in chapter 13.

If you want to go on to the next chapter, leave ArcView running. Otherwise, choose Exit from the File menu.

SECTION 1:
ArcView GIS basics

Classifying and displaying themes

Classifying features based on their attributes

Creating different legends

Using different classifications

Classifying and displaying themes

In chapter 4, you learned the importance of classifying information into groups to make it more meaningful. You also learned the importance of presenting your information effectively. In this chapter, you'll learn how to use the tools in ArcView GIS for classifying and presenting information.

ArcView's Legend Editor gives you a variety of ways to classify and display feature attributes. You can choose a presentation style by selecting the appropriate legend. The type of legend you select determines whether or not the features will be divided into classes and which feature attribute you will use for classification. You can select a classification method, which determines how the attributes are assigned to classes and how many classes there will be. You can also use the Legend Editor to create your own customized classes and displays.

Classifying features based on their attributes

Imagine that you work for an advertising agency. The agency has a prospective client who wants to market a new product in an eighteen-county area. Your department has designed a great, but somewhat expensive, newspaper advertising campaign. Knowing that the client has a limited advertising budget, your boss needs to convince him that running a more expensive campaign in a few counties with high population will get better results than running a cheaper campaign in all the counties. As part of the presentation, your boss asks you to use the agency's desktop GIS, ArcView, to prepare a map of the eighteen-county area showing how the population is distributed.

Exercise 9a

1. If necessary, start ArcView. From the File menu, choose Open Exercise. In the Exercises scrolling list, select "ex9a," then click OK. When the project opens, you see a view with one theme, Counties.

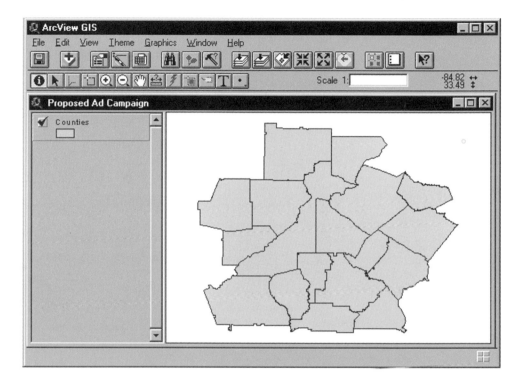

By default, ArcView assigns a single color to all the counties (this is the *Single Symbol* legend type). You want to display the counties according to their population in the attribute table, so you'll open the Legend Editor and design a new legend.

2.   Click the Edit Legend button to display the Legend Editor.

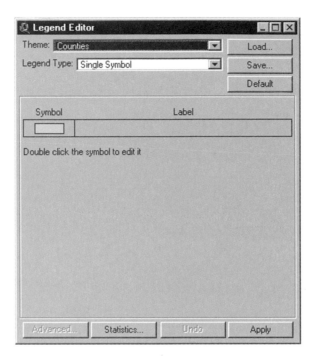

The Theme drop-down list displays the name of the active theme, Counties. The Legend Type list shows that you're using a single symbol for all features. The Symbol column shows the particular symbol being used.

You'll use a different legend to display the counties by their population values.

3.   Click on the Legend Type down arrow and select "Unique Value."

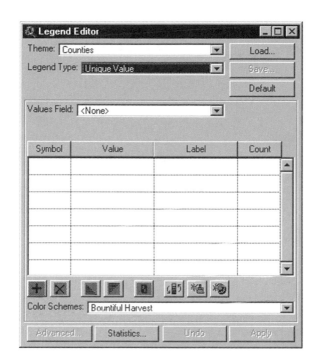

Notice that the look of the Legend Editor has changed; it displays new choices that apply only to the *Unique Value* legend. In this type of legend, each feature attribute with a unique value will be assigned its own symbol in the Symbol field. There are no symbols shown because you haven't yet selected an attribute to symbolize.

4. From the Values Field drop-down list, select "Pop_93" as the attribute whose value will be displayed.

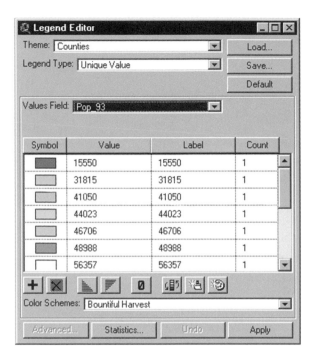

Now you see the symbol for each county, the population value for that county, and the label that will appear in the legend. The Count field tells you how many counties each symbol represents. (If two counties happened to have the same population value, then the Count field would be 2.) The symbol colors are randomly chosen from the colors available in the Bountiful Harvest color scheme. The Color Schemes drop-down list gives you many more choices.

5. Click Apply. If necessary, minimize the Legend Editor or move it out of the way so you can see the view.

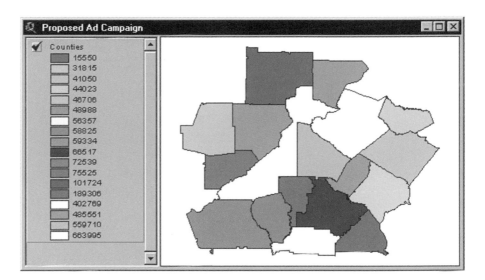

Each county now has its own symbol because each has a unique population value. Looking at this map, you decide that it just won't work for your presentation. It's too hard to tell at a glance which counties are the most populous. But grouping the counties into classes should help.

6. If necessary, restore the Legend Editor. Choose "Graduated Color" from the Legend Type drop-down list.

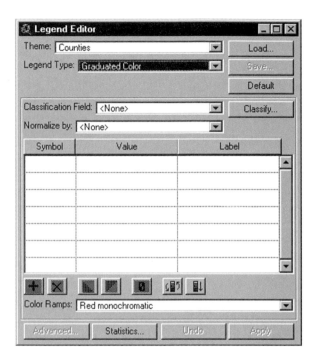

The Legend Editor has changed again. The *Graduated Color* legend requires that you choose an attribute to classify. It will assign colors in graduated shades to the symbols for the classes. No symbols display here because you haven't chosen an attribute in the Classification Field yet.

7. Click on the Classification Field down arrow to display the list of
    attributes for the Counties theme. (Only numeric attributes appear
    because only they can be classified.) Click on "Pop_93" to select it.

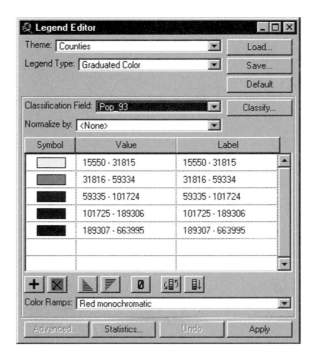

Once you select the Pop_93 attribute, ArcView applies a default classification method, called *Natural Breaks,* to the population values for the
counties. Five classes are created, with class divisions corresponding as
closely as possible to actual breaks in the range of values in the data.
(You'll learn how to change the classification method later in the chapter.)

The attributes in this legend type can also be *normalized* by another
attribute in the Normalize by drop-down list.

Normalizing attributes. When you normalize an attribute, you compare it to another attribute. Normalizing expresses the relationship between two attributes as a ratio. For example, if you normalize population by area, the legend shows population density (for example, people per square mile) instead of population. You could also normalize the population of each county by the total population to display the percentage of the total population that lives in each county. For more information, search for these Help Topics: *Making your data easier to understand, Normalize by another field, Normalize by percent of total.*

The symbols for the classes graduate in color from light red for counties with the lowest population to dark red for counties with the highest population. This is a *color ramp*. The Graduated Color legend uses a color ramp to show increasing class values as darker shades of a color. ArcView offers a variety of color ramps using either one or two colors. Monochromatic color ramps use shades of a single color to display classes. Dichromatic color ramps use shades of two colors to display classes. You can also create your own custom color ramps.

You decide to change the color ramp.

8. Select Green monochromatic from the Color Ramps drop-down list.

9. Click the Apply button to apply the new legend to the Counties theme in the view. If necessary, move the Legend Editor dialog box out of the way or minimize it so you can see the changes in the view.

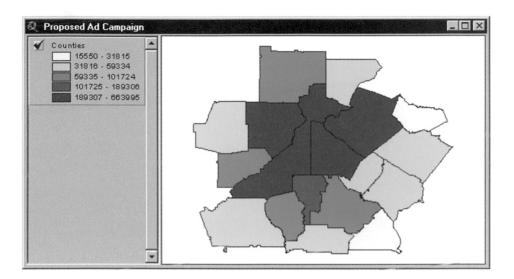

The graduated colors help a lot. The color ramp makes it easy to see the population distribution in the eighteen-county area.

10. Close the Legend Editor.

If you want to continue, leave *ex9a.apr* open.

The legend you select makes a big difference in how features are displayed in your map. Using different legends, you can give your boss several maps to choose from for the presentation.

Creating different legends

So far, you've used the Single Symbol, Unique Value, and Graduated Color legends. ArcView also provides a *Dot* legend type that shows the density of an attribute, and a *Chart* legend type that displays multiple attributes as charts inside each feature. (Another legend type, *Graduated Symbol*, can be used with point and line themes, as you'll see in the next chapter.)

You decide to test the Dot and Chart legend types to see if they might be useful for the presentation.

Exercise 9b

1. If *ex9a.apr* is open, continue. Otherwise, choose Open Exercise from the File menu. In the Exercises scrolling list, select "ex9b," then click OK. You see a view of the Counties theme, classified as it was in the last exercise.

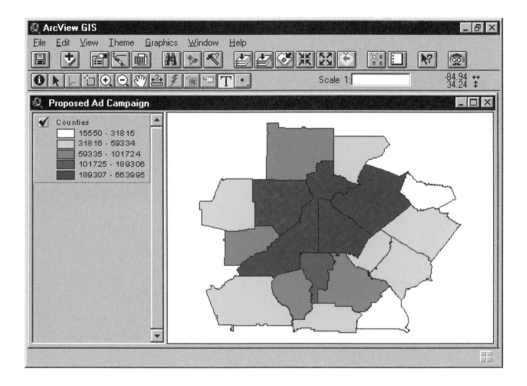

2. Double-click on the Counties legend in the view to open the Legend Editor. In the Legend Type drop-down list, select "Dot."

ENVIRONMENTAL SYSTEMS RESEARCH INSTITUTE, INC.

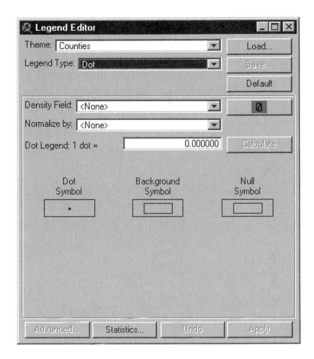

The Legend Editor has changed again; now it has a Density Field, a Dot Legend field, and a Calculate button. It also shows the symbol that will be used for the dots on the map, the background symbol that will be used to fill the polygons, and the null symbol, used to fill polygons that don't have a value for the selected attribute (for example, a county without a population value).

Using dot density maps. You can use dot patterns to show the concentration of an attribute in an area. For example, you could show how densely populated a county is by using one dot to represent every 1,000 inhabitants. If the county has a population of 100,000, it will have 100 dots inside it on the map. A densely populated area will have more dots than a lightly populated area. For more information, search for this Help Topic: *Dot Density Legends*.

3. From the Density Field drop-down list, select "Pop_93" as the attribute whose density will be displayed.

If you know exactly how many people you want each dot to represent, you can type the number into the Dot Legend box, but it's easier to let ArcView calculate the number for you. To do that, ArcView takes into account the dot size (shown in the Dot Symbol box), the display size of the features in the view, and the size of the View window itself.

4. Click the Calculate button. ArcView calculates the number of people each dot will represent. (The value you get will be different than the value shown here if you have changed the size of the view window.)

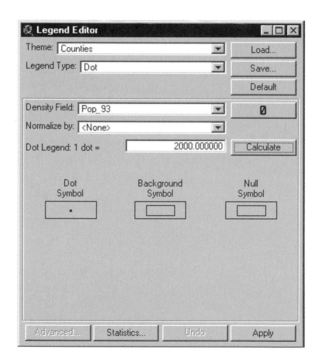

5. Click Apply. ArcView creates a dot density map. If necessary, minimize the Legend Editor or move it out of the way so you can see the view.

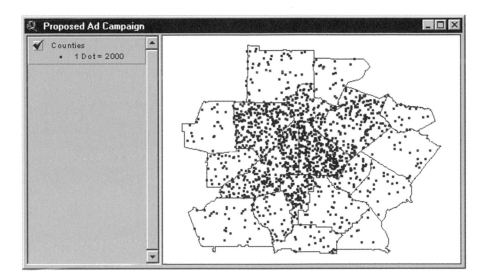

The dot density map shows that the population is denser toward the center of the eighteen-county area.

The ad campaign is targeted at people in their thirties to mid-fifties, so your boss also wants to see the distribution of age groups in the counties. You decide to create a map using the Chart legend type, which will display a bar or pie chart of the selected attributes inside each feature in the view. You'll place a chart symbol in each county showing the distribution of people under 30 years old, 30 to 54, and over 55.

6. In the Legend Editor, select "Chart" from the Legend Type dropdown list. Again, the Legend Editor changes to reflect the new legend type.

First, you'll select the three attributes to be charted, then you'll select the bar chart symbol type.

7. In the Fields column (on the left), choose "Pct_0_29" (the percent-
    age of people between zero and twenty-nine years of age), then click
    the Add button to display this field on the right. Do the same for the
    "Pct_30_54" and "Pct_55+" fields.

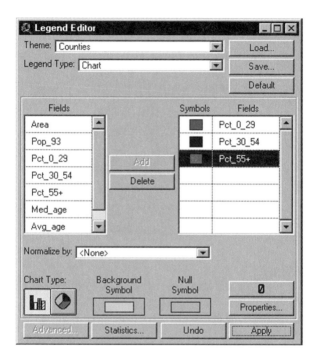

The attributes you add on the right contain the values ArcView will chart.
In the lower left corner of the Legend Editor are two chart buttons, one
for a bar chart and one for a pie chart. You decide to see how the bar chart
looks.

**ArcView randomly assigns a color to each symbol, so your display
may look different.**

8.   Click on the bar chart symbol, then click Apply.

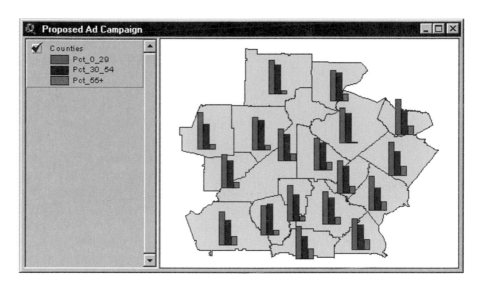

Now the theme shows the age groups represented by a bar chart in each county. Your boss can use this map to show the customer how the age groups are distributed in the eighteen-county area.

9.   Close the Legend Editor.

If you want to continue, leave *ex9b.apr* open.

Using different classifications

You've changed the legend type, but you might also want to change the number of classes or the way the classes are divided. This can change the way the data appears, and hence, how it can be interpreted, so it's important to choose a classification method carefully. The default classification method, Natural Breaks, will divide the counties into five classes. But you decide to simplify things by dividing the counties into high, medium, and low population classes. You could define your high class by including the upper third of the counties ranked by population. Or you could define your high class by a specific number, such as counties with a population over 100,000. A county with a population of 90,000 might be in the high class under the first classification method, but would fall into the medium class under the second method. Because the second method presents the same population value differently, it gives a different message.

> **Choosing a classification method.** You can choose from several different classification methods in the Classification dialog box. All of the classification methods rank the selected attributes from low to high value, then subdivide the ranked attributes into classes. The *Quantile* method creates classes that have the same number of features in each class, and the *Equal Area* method creates classes where the sum of the areas of the features in each class are approximately equal. The *Equal Interval* method creates classes of equal range. The *Standard Deviation* method calculates the mean value and creates classes of equal range above and below it. The default classification method, *Natural Breaks*, creates classes by grouping clusters of similar values in the range of values. You can also create your own custom classifications. For more information, search for this Help Topic: *Classification methods.*

First you'll use the Quantile classification method to divide the eighteen counties into three classes that each have six counties in them.

Exercise 9c

1. If *ex9b.apr* is open, continue. Otherwise, choose Open Exercise from the File menu. In the Exercises scrolling list, select "ex9c," then click OK.

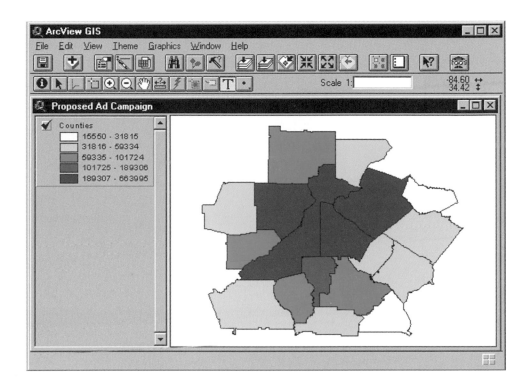

You see the Counties theme displayed using a graduated color legend with a green color ramp. There are five population classes based on the default classification method, Natural Breaks.

2. Double-click on the Counties theme to open the Legend Editor. Click the Classify button. The Classification dialog box displays.

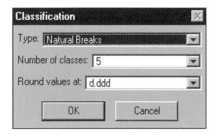

3. In the Classification dialog box, choose "Quantile" from the Type drop-down list and "3" as the number of classes. Click OK.

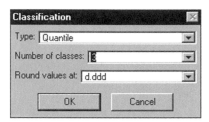

4. In the Legend Editor, the color ramp has reverted to Red monochromatic, so change it back to Green monochromatic in the Color Ramps drop-down list. Click Apply in the Legend Editor. The new classification scheme is applied to the view.

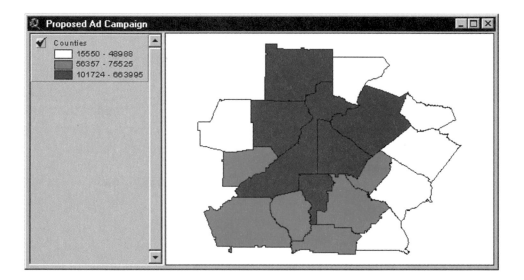

You see that the eighteen counties have been evenly divided into three classes of six counties each. The population range for the classes varies widely, however. It might be better to use the Equal Interval method to divide the range of population values evenly into high, medium, and low classes for the presentation.

5. Click the Classify button in the Legend Editor. The Classification dialog box displays.

ENVIRONMENTAL SYSTEMS RESEARCH INSTITUTE, INC.

6.   Change the Type drop-down list to "Equal Interval" and click OK.

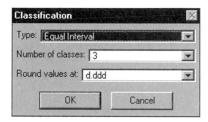

7.   Click Apply in the Legend Editor. Minimize the Legend Editor or move it so you can see the view.

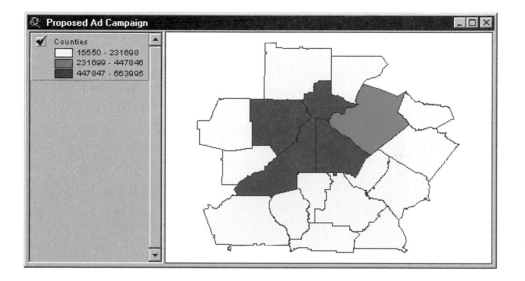

ArcView divided the entire population range of the counties (from 15,550 to 663,995) into three classes that each have an equal range (216,148). The number of counties in each class varies, but the intervals between the high and low values in the classes are the same.

You've tried the Natural Breaks, Quantile, and Equal Interval classification schemes already, and you've seen how different classification methods change the appearance of your data.

At this point, you decide to show the maps to your boss and let her decide which classification she likes best. She likes what you've done so far, but says she knows from experience that it won't be worth running an expensive newspaper advertising campaign in counties with a population of less than 50,000. In fact, the optimum county population for the campaign is over 100,000.

Based on your boss's comments, you decide to create a custom classification. You'll create three population classes: less than 50,000, 50,000–100,000, and greater than 100,000.

8. If necessary, open the Legend Editor. Click on the first value range in the Value column and change it to **0 - 49999** (without a comma), then press Enter. Now change the second value range to **50000 - 100000** and press Enter. Finally, change the third value range to **100001 - 663995** and press Enter. Each time you press Enter, the Label field is updated as well. The new classification looks like this:

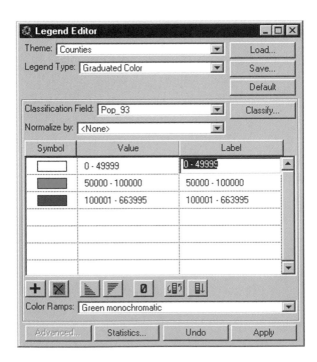

ENVIRONMENTAL SYSTEMS RESEARCH INSTITUTE, INC.

You'll edit the labels to make them a little easier to read.

9. Change the first label in the Label column to **< 50,000** (with a comma) and press Enter. Now change the second label to **50,000 - 100,000** and press Enter. Finally, change the third label to **> 100,000** and press Enter. The new labels look like this:

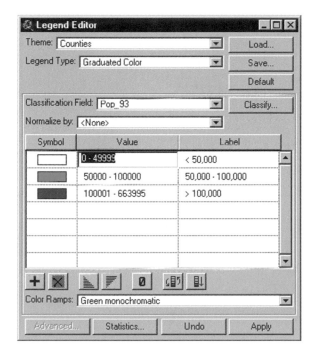

10. Click Apply in the Legend Editor. Your new labels are applied to the Counties theme in the view.

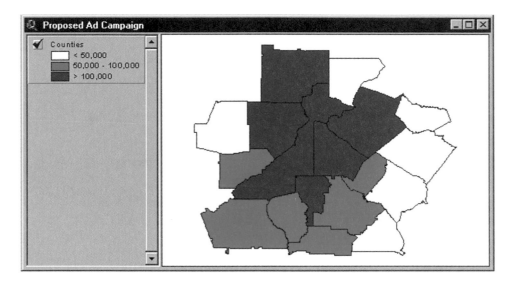

The custom classification clearly shows which counties have populations that are well suited to the expensive ad campaign. Your boss can use this map to show the customer where to concentrate his advertising dollars.

11. Close the Legend Editor.

In this chapter, you've seen how different legends and classification methods affect the way features are classified and displayed. In the next chapter, you'll learn how to use colors, symbols, and labels to complete your presentation.

Symbolizing themes

Using markers and graduated symbols

Using pens, fills, and colors

Using labels and graphics

Symbolizing themes

In chapter 9, you grouped features into classes according to their attributes. Now you'll learn how to apply marker, pen, and fill symbols to each class and how to change the color of the symbols. You'll also learn how to label features and add text to a view.

With the ArcView GIS Legend Editor and Symbol Window, you can choose appropriate symbols to display point, line, and polygon themes. You can also specify the font style, size, and color you'll use to label theme features.

Using markers and graduated symbols

Suppose that you're designing a tourist brochure for the Marsabit National Park and Reserve in northern Kenya. You need a map for the brochure showing the park and reserve areas that tourists might visit, the nearby towns and villages in which they might stay, and the roads and landing strips that provide access to the park. You've loaded the themes for your map into ArcView and now you'll use the Legend Editor and Symbol Window to choose the appropriate colors and symbols for the map.

Exercise 10a

1. If necessary, start ArcView. From the File menu, choose Open Exercise. In the Exercises scrolling list, select "ex10a," then click OK. When the project opens, you see a view of the Marsabit National Park and Reserve.

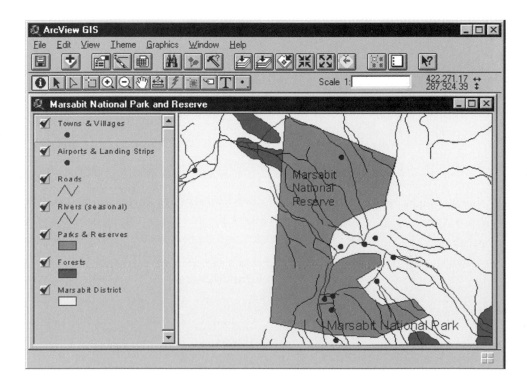

The map shows the park and reserve areas, as well as the surrounding towns and villages, forests, roads, and landing strips, but it's hard to read. First, you'll select a new symbol for the Airports & Landing Strips theme that's appropriate for an airport.

2. Double-click on the Airports & Landing Strips theme in the view's Table of Contents to make the Legend Editor active.

3. Double-click on the symbol for the Airports & Landing Strips in the Legend Editor. The Symbol Window opens to the Marker Palette.

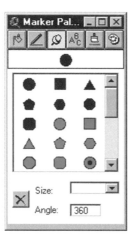

Green markers with black outlines are two-color markers. You can change the color of the green area, but the marker always has a black outline. Black markers are one solid color, which can be changed. The Size drop-down list lets you change the marker size, and the Angle field lets you change the marker rotation. The button with an "X" deletes the selected marker.

Understanding the Symbol Window. ArcView provides several different palettes (collectively called the *Symbol Window*) from which you can change the symbols used to display a theme. You change palettes by clicking on the buttons across the top of the Symbol Window. Use the Fill Palette for polygons; the Pen Palette for line features; the Marker Palette for point features; the Font Palette for text; the Color Palette to assign colors to fills, pens, markers, and fonts; and the Palette Manager to load, save, clear, and create a palette. For more information, search for these Help Topics: *Symbol Window, ArcView Palette Files.*

ENVIRONMENTAL SYSTEMS RESEARCH INSTITUTE, INC.

4.   Scroll through the Marker Palette to view all the available markers.

There doesn't seem to be an ideal marker for the Airports & Landing Strips theme in the default marker palette, so you decide to add a marker palette that contains airplane markers.

Adding markers. There are three ways to add markers to the Marker Palette: you can create markers from fonts, load new marker palettes from a file, or import an image as a marker. To create markers from fonts, you select a font in the Font Palette and click the Create Markers button. The characters in the font become markers in the Marker Palette. ArcView includes fonts that have cartographic symbols as their characters. You can also add markers by loading palette (*.avp*) files from the Palette Manager in the Symbol Window. ArcView comes with additional palette files for markers, lines, fills, and colors. You can also import an image as an individual marker in the Palette Manager. For more information, search for these Help Topics: *Symbol Window, Marker Palette (Dialog box), Palette Manager (Dialog box), Customizing ArcView's symbol palettes.*

Because it's easy to create markers from fonts, you decide to use one of the special cartographic fonts installed with ArcView. These cartographic fonts store symbols as characters in a font file outside of ArcView until you actually need them. That way, the Symbol Window loads faster and isn't cluttered with extra markers. The ESRI Transportation & Municipal font has an airplane marker you can use for the Airports & Landing Strips theme.

5.   Click the Font Palette button at the top of the Symbol Window. Widen the Font Palette so you can see the entire name of each font.

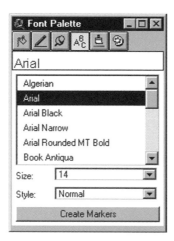

The default font in ArcView is Arial. (The fonts you see listed will depend on which fonts you have loaded on your computer.) The Size drop-down list shows the size of each character in *points,* where one point equals 1/72 of an inch. The Style drop-down list shows the font style (Normal, Italic, Bold, or Bold Italic).

6.   Scroll down to the ESRI Transportation & Municipal font and highlight it by clicking on it. Click the Create Markers button at the bottom of the Font Palette.

The Symbol Window changes to the Marker Palette after you press Create Markers. The ESRI Transportation & Municipal font characters become markers at the bottom of the Marker Palette.

7.   Scroll down and select the airplane symbol by clicking on it. In the Size drop-down list, select 12.

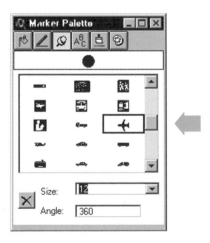

Now the Airplane symbol will be 12 points high in the view.

8.   Click Apply in the Legend Editor. If necessary, move or minimize the Legend Editor and Symbol Window so you can see the view.

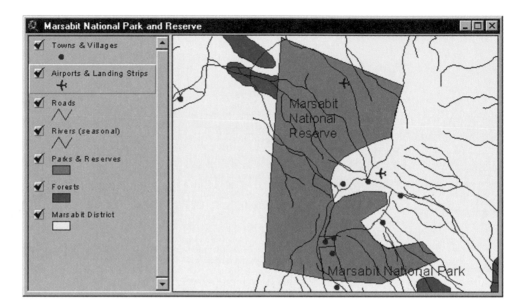

The airplane marker makes it clear which points are airports and landing strips.

Now you're going to change the symbol for the Roads theme to distinguish three grades of roads: international, primary, and minor unpaved. It's important to show the road grades in the brochure so visitors will realize that they may need four-wheel-drive vehicles. The Graduated Symbol legend type has symbols that increase in size as the value of the attribute being classified increases. You'll use it to show major roads as thicker lines and minor roads as thinner lines.

9.   Double-click on the Roads theme in the view's Table of Contents. The Legend Editor for the Roads theme opens.

10. From the Legend Type drop-down list, select "Graduated Symbol." In the Classification Field drop-down list, select "Road_Code."

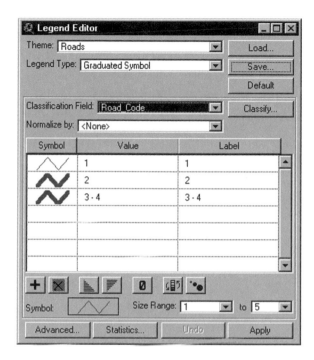

The Legend Editor shows the international (class 1) roads using the thinnest line width, primary (class 2) roads as a thicker line, and minor unpaved (class 3) roads as the thickest line. You want to reverse the graduation of the symbols so that international roads have the thickest line width and minor unpaved roads have the thinnest.

11. Click the Reverse Symbols button at the bottom of the Legend Editor.

Now the international roads are thicker than the primary and minor unpaved roads.

Next you'll change the labels for the road classes so they are more descriptive than numeric codes.

12. In the Label field, type **International** for class 1, **Primary** for class 2, and **Minor Unpaved** for classes 3–4.

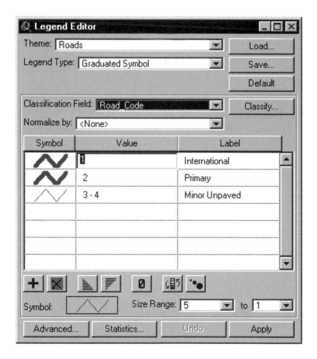

13. Click Apply in the Legend Editor. If necessary, minimize the Legend Editor or move it out of the way so you can see the view.

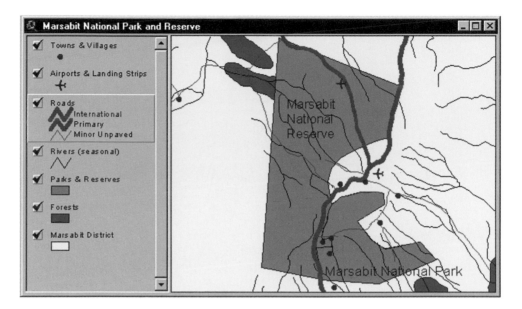

This map shows visitors that the airports and landing strips are next to major paved roads. It also shows which roads are unpaved and may require a four-wheel-drive vehicle.

If you want to go on to the next exercise, leave the project open.

Using pens, fills, and colors

Your map is much easier to read, but you still want to touch it up for the brochure. You decide to change the symbol for the Rivers theme to a dashed line because the rivers are seasonal. You also want to show all of the Forests theme, so you'll change the fill for the Parks & Reserves theme to a pattern with a transparent background.

Exercise 10b

1. If *ex10a.apr* is open, continue. Otherwise, choose Open Exercise from the File menu. In the Exercises scrolling list, select "ex10b," then click OK. You see a view of the Marsabit National Park and Reserve with themes classified as they were in the last exercise.

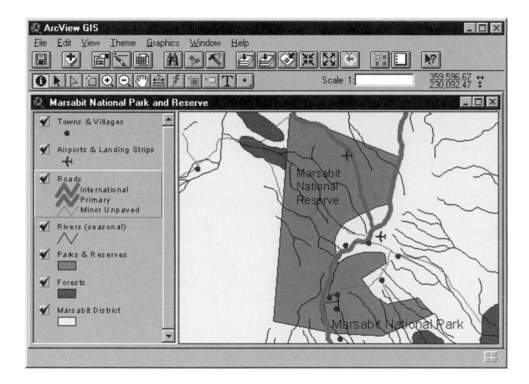

2. Double-click on the Rivers theme in the view's Table of Contents to open the Legend Editor.

3. Double-click on the Rivers symbol in the Legend Editor. The Symbol Window opens to the Pen Palette.

4.   Select the dashed line shown below for the Rivers symbol.

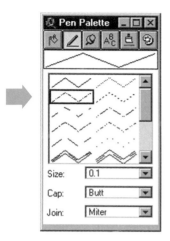

The Size drop-down list lets you select the line width. The Cap drop-down list defines how the ends of lines will appear (Butt, Square, or Round). The Join drop-down list defines how two line segments should meet at a vertex (Miter, Round, or Bevel).

5.   Click Apply in the Legend Editor. If necessary, minimize the Legend Editor or move it out of the way so you can see the view.

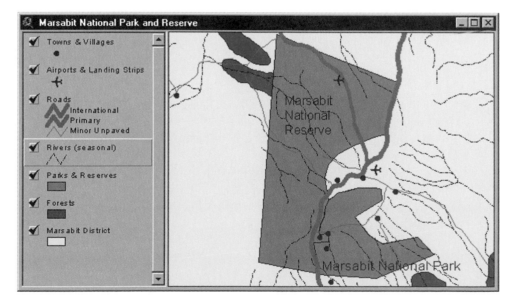

Now the rivers appear as dashed lines in the view, indicating that they are seasonal. Next, you'll change the Parks & Reserves symbol to a pattern with a transparent background, so that all of the Forests theme will show.

6. Double-click on the Parks & Reserves theme in the view's Table of Contents. The Legend Editor for the Parks & Reserves theme opens.

7. Double-click on the symbol for the Parks & Reserves theme in the Legend Editor. The Symbol Window opens to the Fill Palette.

8. Select the fill shown by clicking on it.

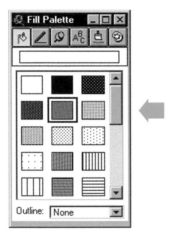

The Fill Palette has choices for transparent, solid, or patterned fills. You've selected a pattern, but the default background color is opaque white. You'll change it to transparent so that the Forest theme can be seen beneath the Parks & Reserves theme.

9. Click the Color Palette button to open the Color Palette.

The Color drop-down list in the Color Palette gives you four choices: Foreground, Background, Outline, and Text.

10. Choose Background from the Color drop-down list and select Transparent (the box with an "X" through it in the upper left corner).

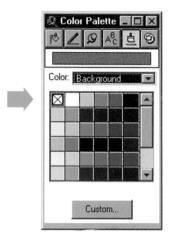

The Color Palettc has several colors in many shades. The Custom button allows you to define others.

11. Click Apply in the Legend Editor. If necessary, minimize or move the Legend Editor so you can see the view.

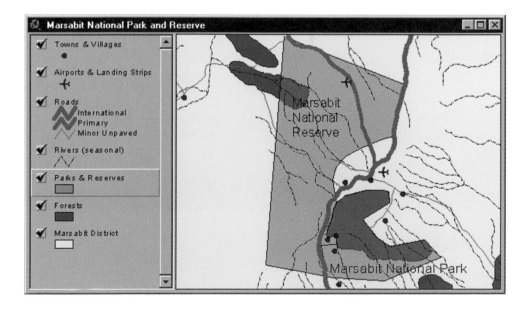

You can now see the forest areas inside the park boundaries that tourists may want to visit.

If you want to go on to the next exercise, leave the project open.

Using labels and graphics

You've chosen appropriate symbols for your map, but you still need to label the towns, villages, and roads by name. Inside your brochure, you list the towns and villages that have visitor accommodations, so it's important to label them on the map, too.

Labels are a special kind of graphic that you create from attributes of a theme. You can select an attribute in the Theme Properties and use Auto-label or the Label tool to display the names of features in the map. For the Towns & Villages theme, you'll select the Name attribute for labeling.

With the Text tool, you can create text that appears in the view but is not associated with a theme. Since the attribute table of the Roads theme doesn't have a field that would make a good label, you'll use the Text tool to add text identifying the major roads on the map.

Exercise 10c

1. If *ex10a.apr* or *ex10b.apr* is open, continue. Otherwise, choose Open Exercise from the File menu. In the Exercises scrolling list, select "ex10c," then click OK. You see a view of the Marsabit National Park and Reserve with themes classified as they were in the last exercise.

ENVIRONMENTAL SYSTEMS RESEARCH INSTITUTE, INC.

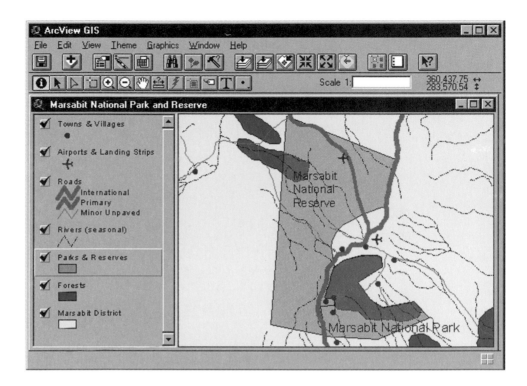

You'll auto-label all of the towns and villages at once, but first you'll select a font to use for the labels. (If *ex10a.apr* is still open, the ESRI Transportation & Municipal font is still selected. It has symbols instead of characters.)

2.   Make the Towns & Villages theme active by clicking once on it in the view's Table of Contents.

3.   From the Window menu, choose Show Symbol Window. The Symbol Window opens.

4. Click the Font Palette button, then click on the Arial font (if it's not already selected). In the Size drop-down list, select 12. (Text size is measured in points, the same as marker size.)

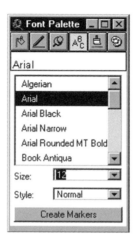

Now you can label all of the points at once using ArcView's Auto-label feature. (You could label the points individually by clicking on each one with the Label tool, but when there are several points, Auto-label is easier.)

5. From the Theme menu, choose Auto-label. The Auto-label dialog box displays.

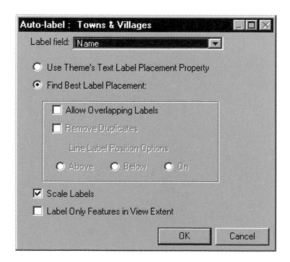

The Label field drop-down list shows that the Name attribute will be used for labeling. The Find Best Label Placement check box is checked, so ArcView will calculate the best location for each label. The Scale Labels check box is also checked, so the labels will change size if you zoom in or out.

6.   Click OK. ArcView labels the towns and villages (the active theme) with their names.

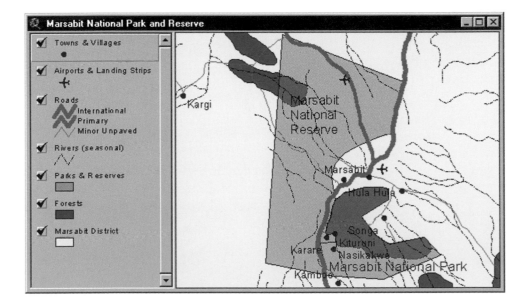

Next, you'll label the two major roads going into the park and reserve.

The attribute table for the Roads theme doesn't include a field for the road name, so neither Auto-label nor the Label tool will be useful in this case. Instead, you'll create text using the Text tool.

7.   Click on the Text tool, then click near the international road (the thickest road symbol running vertically through the center of the view) where you want to put the road name. The Text Properties dialog box displays.

The international road is called "A2."

8.   Type **A2** in the Text Properties dialog box and click OK.

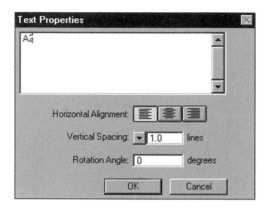

9.   Click on the Pointer tool and move it over the text until the cursor changes to a four-headed arrow. Drag the text to place it exactly where you want it, next to the international road.

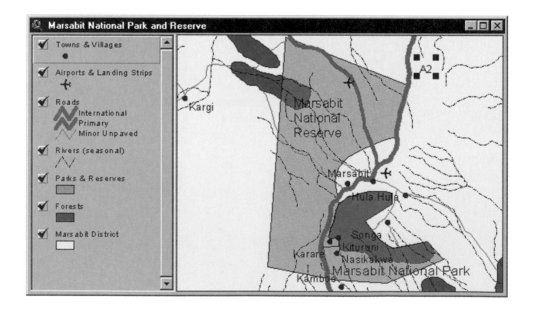

10. Repeat steps 7–9 to put the name "C82" next to the primary road (the left arm of the "Y" formed by the major roads).

11. Click anywhere away from the text graphic to make the selection handles disappear.

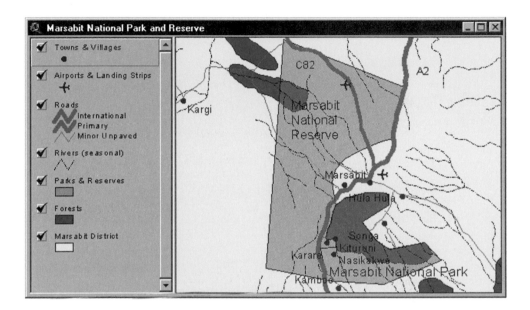

12. Close the Font Palette.

The map is ready for the brochure. If you compare this version to the map you started with, you can see the improvement that well-chosen symbols make. The map is more attractive and, more importantly, it's much more informative. Tourists can locate the towns and villages by name to make accommodations. They can locate the forested areas they may want to visit. And they can easily find the airports and landing strips and see the quality of roads they'll travel on.

If you want to go on to the next chapter, leave ArcView running. Otherwise, choose Exit from the File menu.

SECTION 2

Working with
spatial data

The next two chapters introduce you to how to reference spatial data to the real world. In chapter 11, you'll use the measuring tools in ArcView GIS to measure distance and area in a view. You'll also change the map projection used to display a view and witness the effects that different map projections have on distance and shape. In chapter 12, you'll learn how to set the map scale for a view and how to set scale thresholds for individual themes so you can control when they display.

SECTION 2:
Working with spatial data

Measuring distance and area in a view

Measuring distance

Measuring area

Setting a map projection

Working with data that's already projected

CHAPTER 11

Measuring distance and area in a view

In chapter 3, you learned that one of the advantages of a desktop GIS is that you can measure distances and areas on the surface of the earth right at your computer. And the measurements are reported in real-world units, such as feet, meters, miles, or kilometers. You also learned that you can control the distortion in certain properties by using a map projection that preserves the property you're measuring.

In ArcView GIS, you can measure distance and area in a view whether the feature locations are stored as unprojected geographic coordinates or as projected x,y coordinates. All you have to do is tell ArcView what units the coordinates are stored in and what units you want ArcView to use for reporting measurements.

In this chapter, you'll learn how to measure distance and area when the feature locations are unprojected (stored in geographic coordinates), when you set a map projection for the view, and when you use data that's already been projected.

Measuring distance

You work for a city that's planning to develop a park adjacent to a soon-to-be-developed housing project. The park property has one building on it that will be converted to a rest room. It's your job to make some preliminary measurements that will help the city estimate the cost of developing the park. You'll first consider the cost of bringing water into the park's future rest room, then make some measurements to help determine this cost.

Exercise 11a

1. From the File menu, choose Open Exercise. In the Exercises scrolling list, select "ex11a," then click OK. When the project opens, you see a view containing three themes: Water Lines, Buildings, and Property. All of these themes contain geographic data that's stored in decimal degrees. (*Decimal degrees* are degrees of latitude and longitude expressed as a decimal rather than as degrees, minutes, and seconds.)

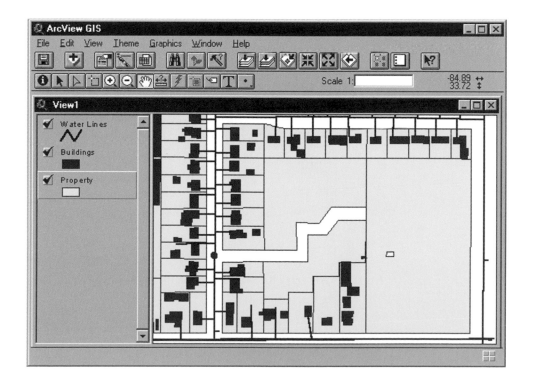

Notice that the Scale box on the tool bar is empty. That's because you haven't told ArcView what units to use. Before you make any measurements in the view, you'll tell ArcView what type of units the data is stored in and what type of units you want to use for measuring. Then ArcView will be able to calculate the map scale.

If you don't know the units your data is stored in, you may need to consult the data dictionary (if your data comes with one), the vendor (if you purchased the data), or the person or agency you received the data from. If the data is an ARC/INFO coverage or grid, it may contain a text file (prj) that describes the coordinate system and units the data is stored in.

2. Select Properties from the View menu to open the View Properties dialog box.

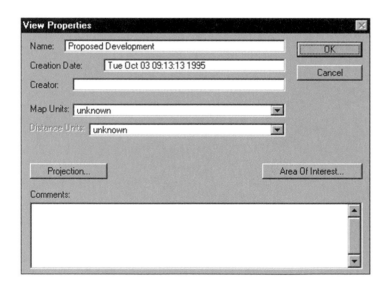

To specify the units in which the coordinates of the data are stored, you'll set the map units.

3. Click on the Map Units down arrow, then choose "decimal degrees" from the list. This indicates that all the data in the current view is stored in decimal degrees. To specify the units ArcView will use to report measurements, you'll set the distance units.

4. Click on the Distance Units down arrow, then choose "feet."

Before returning to the view, you decide to give it a more descriptive name.

5. Click in the Name field and highlight "View1." Change the name to **Proposed Development.**

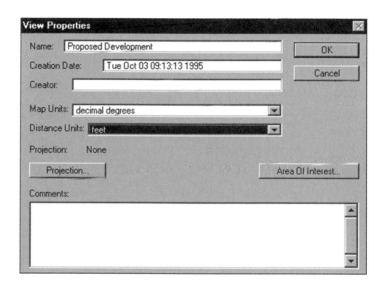

6. Click OK to apply your settings to the view. Now the view has a new name and the map scale displays in the Scale box.

You'll use the Measure tool to determine the distance from the small building (located on the proposed park property) to the nearest water lines.

7. Click on the Measure tool (the cursor changes to a ruler), then click on the building (highlighted in yellow). Move the cursor to the water line located directly below the building. Notice that ArcView draws a

line segment from the building to wherever you position the cursor in the view and reports the length of this line in the status bar (at the bottom of the ArcView window). With the cursor directly over the water line, double-click the mouse button to end the line. ArcView reports the measurement in feet.

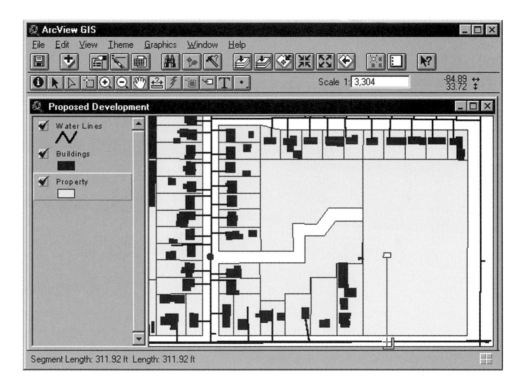

ArcView reports two values, Length and Segment Length. Segment Length is the length of the current line segment and Length is the total length of all segments that comprise the line. In this case, both measurements are the same, 312 feet. (In this exercise, your measurements may vary slightly from those reported.)

8.   Click on the same building again, then move the cursor to the water line located to the far right of the building. With the cursor directly over the water line, double-click the mouse button to end the line. ArcView reports the measurement in the status bar.

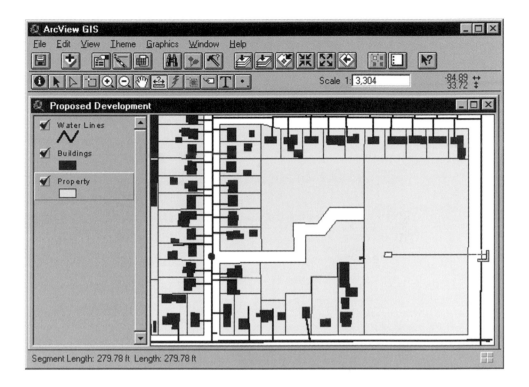

In only a few seconds, and without having to go to the site, you've determined that the water line to the right of the building is approximately 32 feet closer (280 feet) than the water line directly below it (312 feet). You can use this information in calculating the cost of bringing water to the building.

In addition to the proposed park, the city is in the process of approving a housing development on the adjacent property. Before the planning commission gives its final stamp of approval, it needs to estimate the cost of running new water lines into the proposed development.

Because it's common practice for utility companies to use existing rights-of-way to gain access to properties and structures, and the proposed housing development already has a road, you'll use the existing street right-of-way for measuring the new water lines.

9. With the Measure tool selected, click on the red dot located on the street centerline on the left side of the view (this dot indicates where the new water lines will begin).

Now you'll trace the centerline of the street located to the right of the red dot.

10. Move the cursor along the center of the street and single-click the mouse button each time the street changes direction. (Each time you single-click, ArcView reports a measurement and starts a new line segment.) At the end of the street, double-click to end the line.

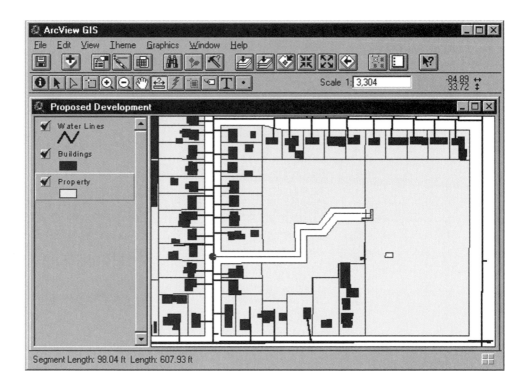

In the status bar, ArcView reports the length of the last line segment (98 feet) and the total length of all the segments (608 feet). Now you know that the length of pipe needed to bring water to the proposed housing development is about 608 feet. You've used ArcView to measure a distance on your computer instead of in the field, saving hours of labor. Imagine the savings if you needed to run several water lines. You could measure them and estimate their cost in minutes rather than hours.

If you want to go on to the next exercise, leave the project open.

Measuring area

A soccer field is planned for the new park. You're fairly certain it will fit, but you don't want to spend hours at the site trying to determine all the possible locations for it. You'll use ArcView to determine the best locations for a soccer field.

Exercise 11b

1. If *ex11a.apr* is open, continue. Otherwise, choose Open Exercise from the File menu. In the Exercises scrolling list, select "ex11b," then click OK. When the project opens, you see a view with three themes: Water Lines, Buildings, and Property.

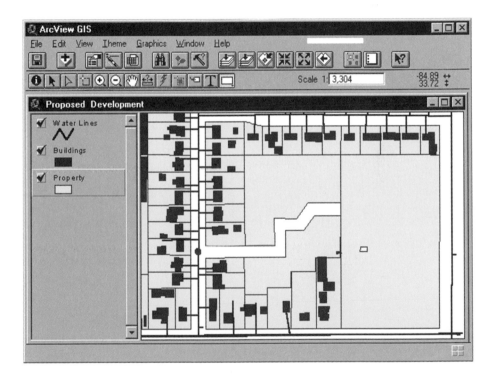

Because a soccer field is measured in meters instead of feet, you need to change the units ArcView uses to report measurements (distance units) to meters.

2. Select Properties from the View menu to open the View Properties dialog box.

3. Click on the Distance Units down arrow, then choose "meters."

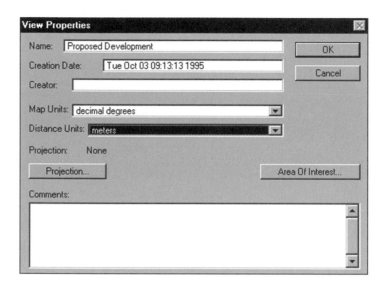

4. Click OK. Your measurements will now display in meters.

You'll use the Rectangle tool to draw the shape of a soccer field, 110 meters by 73 meters.

5. Click on the Draw tool and select the Rectangle tool from the drop-down list of tools.

6. Click inside the proposed park property, hold down the mouse button, and drag a rectangle that measures approximately 110 meters by 73 meters. The status bar indicates the Extent (width, height) and Area of the rectangle as you draw it. When you're satisfied with the rectangle (your measurements don't have to be exact), release the mouse button.

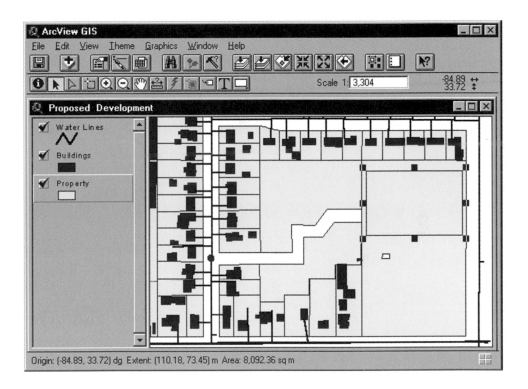

Next you'll use Size and Position to set the exact measurements for the rectangle.

7.   Choose Size and Position from the Graphics menu. The Graphic Size and Position dialog box displays.

8.   In the dialog box, set the width to **110** (meters) and the height to **73.**

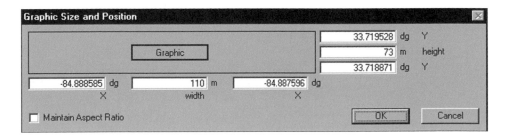

9. Click OK. Now the rectangle has the right dimensions for a soccer field.

To reposition the rectangle, you'll use the Pointer tool.

10. Click on the Pointer tool, then position the cursor over the rectangle (it changes to a four-headed arrow). Now hold down the mouse button and drag the rectangle to any position on the property. In this way you can determine the best locations for the soccer field.

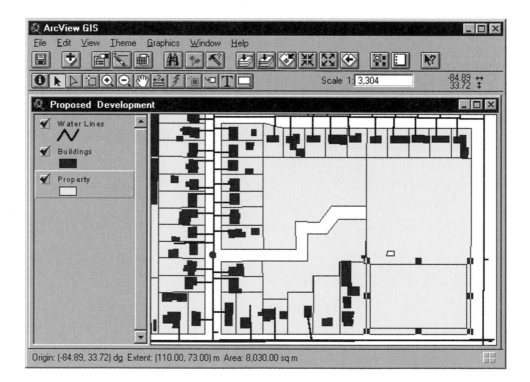

But suppose you want to change the orientation of the soccer field. No problem. Just use Size and Position to change the rectangle's dimensions.

11. From the Graphics menu, choose Size and Position. Change the width to **73** meters and the height to **110** meters, then click OK.

Now you can use the Pointer tool to reposition the rectangle.

12. Place the cursor over the rectangle, then hold down the mouse button and drag the rectangle to any position inside the park property.

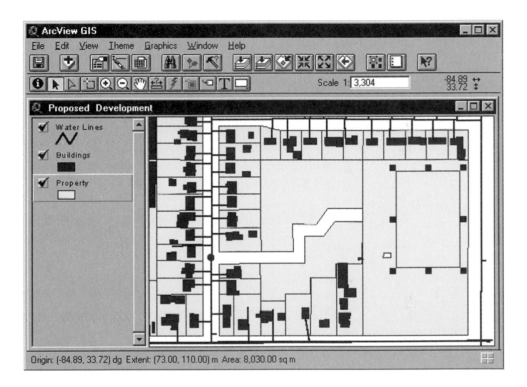

There are many suitable locations for the soccer field. To determine them in the field would take two people, a surveyor's chain or transit, and many hours. ArcView can help you find the possible locations, on your computer, in only a few minutes.

Setting a map projection

You've seen how ArcView lets you work with data that's stored in geographic coordinates without setting a map projection. All you have to do is set the map and distance units for the view. However, when you're working with data that covers a large portion of the earth's surface and you want to preserve a particular spatial property, such as shape, area, distance, or direction, you can choose a map projection that preserves that property and apply it to the view. ArcView provides a wide range of standard and custom map projections to choose from.

You can project a view in ArcView whenever the coordinates of the spatial data (feature locations) are stored in decimal degrees. In other words, the spatial data must be unprojected. You can convert unprojected data from decimal degrees to planar (x,y) coordinates using any projection that ArcView supports.

Suppose you teach a map-reading class at a university. Your lesson for this week is "Understanding Projections." You plan to use ArcView to show students how different projections affect shape and distance. You'll use familiar data (the continental United States) to create a demonstration that shows the changes produced by each projection. To show the distortion of shape, you'll add a circle graphic to the data, then observe how the shape of the circle changes with each projection that's applied. To show the distortion of distance, you'll measure a known distance (the distance from Los Angeles to New York) and compare your measurements.

ENVIRONMENTAL SYSTEMS RESEARCH INSTITUTE, INC.

Exercise 11c

1. From the File menu, choose Open Exercise. In the Exercises scrolling list, select "ex11c," then click OK. When the project opens, you see a view with two themes, Cities and 48 States.

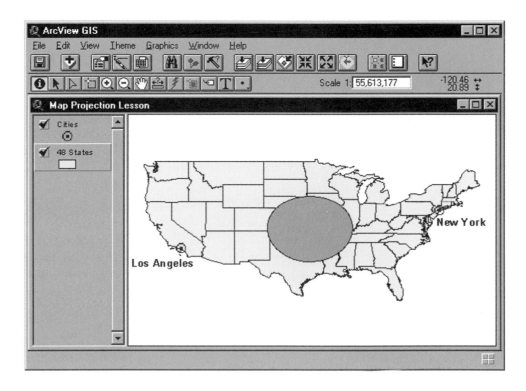

First you'll see how the map units and distance units are set for this view.

2.   From the View menu, choose Properties. The View Properties dialog box displays.

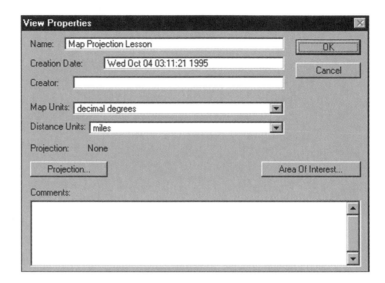

The Map Units are set to "decimal degrees" and the Distance Units are set to "miles," that is, the coordinates of the data in the view are stored in decimal degrees and the measuring units are miles. The projection is set to "None."

3.   Click the Cancel button to close the View Properties dialog box.

Now you'll measure the distance from Los Angeles to New York, which is known to be 2,451 miles.

4.   Click on the Measure tool, then click on the symbol for Los Angeles. Move the cursor to the symbol for New York, then double-click to end the line. The distance between the two cities displays in the status bar.

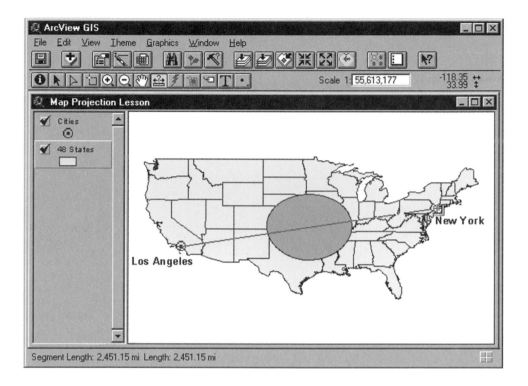

The distance is approximately 2,451 miles (your measurement may vary). Even though the shapes of the individual states are distorted (note the elliptical distortion of the circle), the distance measurement is accurate. That's because, when no projection is set, ArcView computes distance from the spherical coordinates of latitude and longitude. In other words, it's taking the earth's round surface into account.

Now you'll choose a projection for the view and see how it affects distance and shape.

5.   Select Properties from the View menu, then click the Projection button. The Projection Properties dialog box displays.

6.   Click on the Type down arrow, then choose "Mercator" from the list.

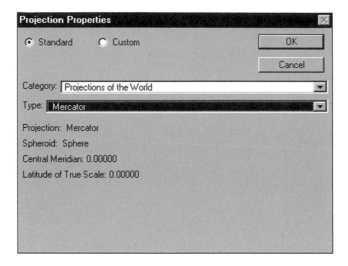

7.   Click OK to return to the View Properties dialog box.

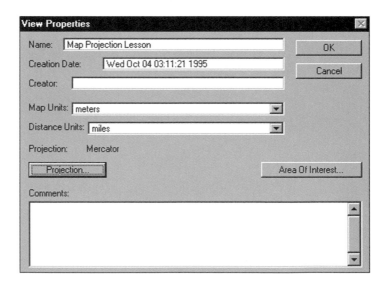

The Projection field indicates that the Mercator projection has been selected.

8. Click OK in the View Properties dialog box to apply the Mercator map projection to the view.

9. With the Measure tool, measure the distance from Los Angeles to New York.

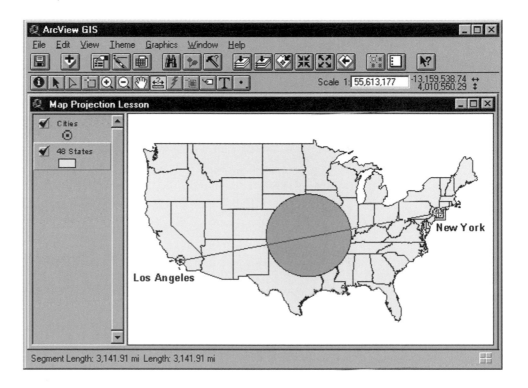

The new measurement, approximately 3,142 miles, is larger than the previous measurement, by about 691 miles.

The ellipse is now a circle (its true shape). The Mercator map projection preserves the property of direction and the shape of features, but sacrifices accurate distance and area.

You'll change the projection again.

10. Select Properties from the View menu and click the Projection button.

11. In the Projection Properties dialog box, select "Peters" from the Type drop-down list.

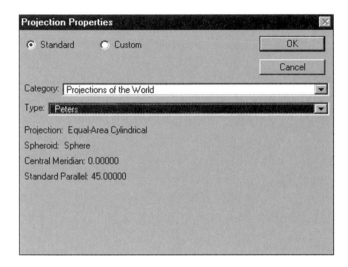

12. Click OK, then click OK in the View Properties dialog box to apply the Peters map projection.

13. With the Measure tool, measure the distance between the two cities.

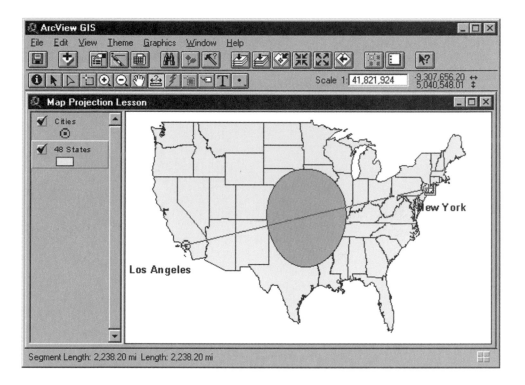

The new measurement (approximately 2,238 miles) is smaller than the original measurement (2,451 miles), by about 213 miles.

The circle is now egg-shaped. The Peters projection preserves accurate area but sacrifices the properties of shape, distance, and direction.

The two previous projections, Peters and Mercator, are most suitable for regions near the equator. Next you'll use a projection that's suitable for the continental United States.

14. From the View menu, choose Properties, then click the Projection button to display the Projection Properties dialog box. Click on the Category down arrow, then select "Projections of the United States."

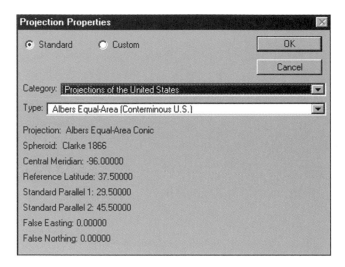

15. Click on the Type down arrow to display a list of projections for the United States, then select "Equidistant Conic (Conterminous U.S.)."

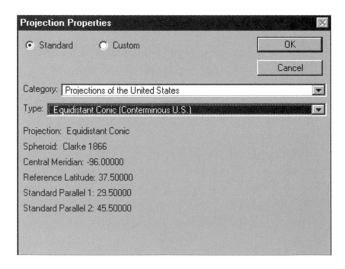

16. Click OK, then click OK in the View Properties dialog box to apply the new projection.

Now you'll measure the distance between Los Angeles and New York again.

17. Click on the Measure tool and measure the distance between Los Angeles and New York.

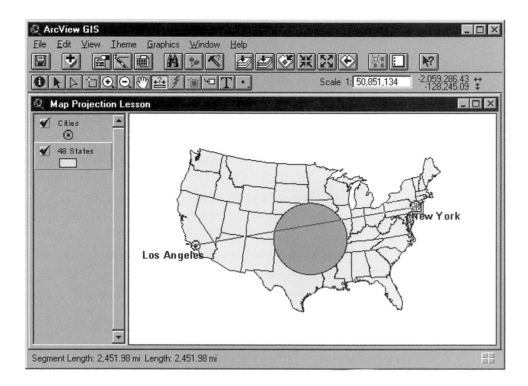

This time, the measured distance (2,452 miles) is almost the same as the unprojected measurement (2,451 miles) in step 4.

The ellipse is now a circle. The Equidistant Conic projection preserves shape and maintains accurate distance in the East–West direction for the lower forty-eight states. However, these qualities are achieved at the expense of direction and area.

With this ArcView demonstration, your students will be able to see how shape and distance change from one projection to another.

> **Learning more about projections.** ArcView provides lists of standard and custom projections. From these lists, you can access more detailed information about any of the projections ArcView supports. For more information, search for these Help Topics: *Map projections, Setting the map projection.*

Working with data that's already projected

Much of the data you buy is stored in geographic coordinates, usually decimal degrees. As you've seen in this chapter, you can work with this data without projecting it, or you can apply any of the standard or custom projections ArcView supports.

If you use data from a governmental or international agency, chances are this data is stored in a projection. You can successfully work with this data in ArcView if you remember a few things. First, you can't change the projection of this data or "unproject" it by converting it to geographic coordinates. Second, before you can perform measurements on this data or display its scale (see chapter 12), you must go to the View Properties dialog box and specify the units the data is stored in. That's because ArcView has no way of knowing what these units are until you specify them.

ArcView applies projections only to feature data, not to image data. If you are working with projected image themes and unprojected feature themes in the same view, you must apply the image themes' projection to the view. ArcView will project the feature themes and leave the image themes alone, so all themes will align. If your feature and image data is already in the same projection, you don't need to set a projection.

Finally, you can't display a mixture of projected and unprojected feature themes, or feature themes with different projections, in the same view. That's because ArcView applies a projection to the entire view, not to individual themes (except image themes, which ArcView leaves alone). To work with feature data in different projections successfully, you must store each theme in a separate view.

If you want to go on to the next chapter, leave ArcView running. Otherwise, choose Exit from the File menu.

ENVIRONMENTAL SYSTEMS RESEARCH INSTITUTE, INC.

SECTION 2:
Working with spatial data

Managing scale

Changing the scale of a view

Setting scale thresholds for themes

Managing scale

Each time you zoom in and out in a view or resize a view's window, its scale automatically changes. ArcView GIS reports the scale in a box located on the right side of the View tool bar. You can set a scale for the view by entering a value in this box. ArcView redraws the view at the new scale. For ArcView to calculate the scale correctly, all you need to do is specify the units the data is stored in.

You can also control the scale at which a theme displays by setting a scale threshold for it. In this way, you can specify that a theme will display only when you zoom in or out to a certain scale. In this chapter you'll learn how to set a scale for a view and how to set scale thresholds for themes.

Changing the scale of a view

Your company is interested in opening a new international sales office in Italy. As the European market analyst for your firm, you've been asked to study potential sites, among them Milano, Genova, Roma, Napoli, and Palermo. After much research, you've concluded that Milano is the best site for the new office. You want to present your arguments to upper management. Some of them aren't familiar with your market region, so you want to use maps at the beginning of your presentation to orient them to the European sales region and the potential sites you considered. Then you want to zoom in on Milano as you describe why it's the best site for the new international sales office. For your presentation, you'll use several of the World data themes that come with ArcView. With these themes and ArcView's zooming tools, you'll create the various scenes for your presentation.

Exercise 12a

1.   If necessary, start ArcView. From the File menu, choose Open Exercise. In the Exercises scrolling list, select "ex12a," then click OK. When the project opens, you see a view of the world from space showing the European sales region, the country of Italy, and the proposed sites (not distinguishable at this scale).

During this exercise, don't maximize or resize the view window, since this alters the view's scale. Your scale values may vary slightly from those reported, depending on your screen's resolution, font size, and other factors.

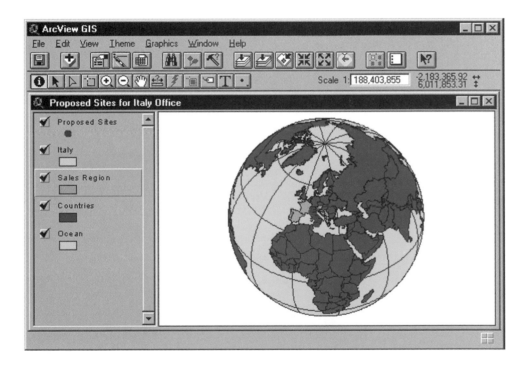

This is the scene you want to show at the beginning of your presentation. Notice that the Scale box on the right-hand side of the View tool bar shows a value of 1:188,403,855. This value is the *scale fraction* or *scale*

ratio. It tells the relationship between the size of the area in the view and the size of the same area in the real world.

> **Understanding scale.** The number that ArcView displays in the Scale box describes the relationship between the dimensions of the view and the dimensions of the earth. For example, if the Scale box displays the number 100,000, then one unit of measurement on the view equals 100,000 of the same unit on the earth, or the features you see in the view are 100,000 times smaller than they are in the real world. The scale is commonly expressed as a fraction (1/100,000) or as a ratio (1:100,000). For more information, search for these Help Topics: *Map scale and accuracy, Setting view scale.*

ArcView calculates a view's scale using the current map units. To determine how the map units for this view are set, you'll display the View Properties dialog box.

2.   From the View menu, select Properties. The View Properties dialog box displays.

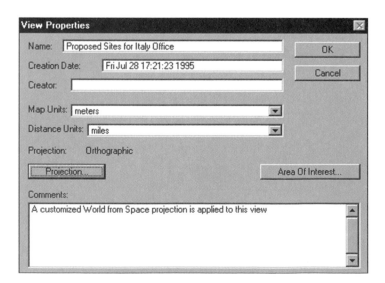

The map units are set to "meters" and a map projection is applied to the view. The Comments section tells you that the projection is customized. (See chapter 11 for a discussion of map projections and map units.)

3.   Click the Cancel button to dismiss the View Properties dialog box.

For your next scene, you want to show a more detailed view of the European sales region, so you'll zoom in to that theme.

 4.   With the Sales Region theme active, click the Zoom to Active Theme(s) button. ArcView zooms in so that features in the Sales Region theme fill the view window.

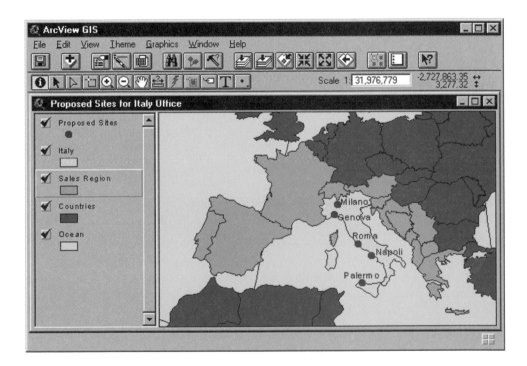

Note that the value in the Scale box changes to 1:31,976,779. This is the scale ArcView calculated based on the size of the view window and the size of the real-world area shown. Notice that the denominator of the scale fraction, 31,976,779, becomes smaller as you zoom in.

For the next scene, you want to zoom in to the Italy theme to look more closely at the proposed office sites.

 5. Make the Italy theme active and click the Zoom to Active Theme(s) button. ArcView zooms so that features in the Italy theme fill the view window. The value in the Scale box changes to 18,407,341.

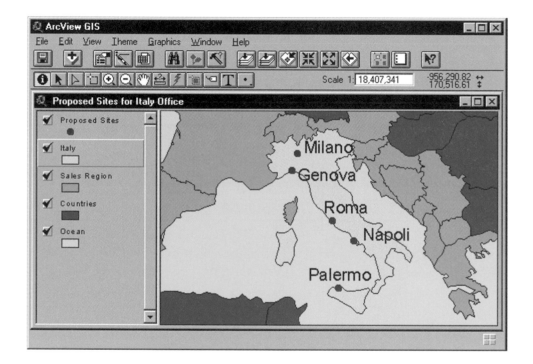

When you present your arguments for siting the new sales office in Milano, you want to show Milano at the center of the view. So, you'll select it, then zoom in to it.

 6. Make the Proposed Sites theme active. Click on the Select Feature tool, then click on the feature (point) that represents Milano. ArcView selects and highlights it.

7. Click the Zoom to Selected button. ArcView pans the view so Milano appears in the center of it. In this case, ArcView doesn't zoom in any further so the scale remains the same.

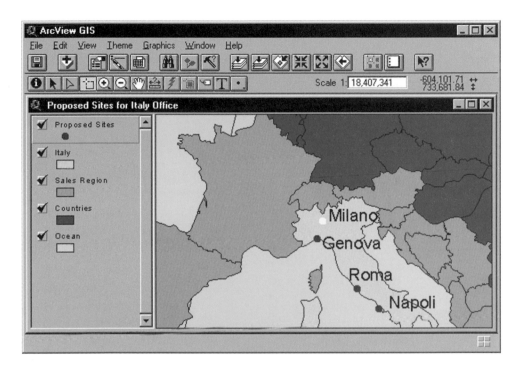

Zooming to the previous extent. There may be times when you have zoomed in or out and you want to return to the previous view extent. The Zoom to Previous Extent button can undo the last five zooms. For more information, search for this Help Topic: *Zoom Previous Extent.*

In step 4, you zoomed in to show all the features in the Sales Region theme. The value in the Scale box was 31,976,779. You'll use this value (rounded off) to set the scale directly.

8. Click in the Scale box, drag the cursor to highlight the current value, then type **32,000,000** and press the Enter key on your keyboard. The view redraws at this scale.

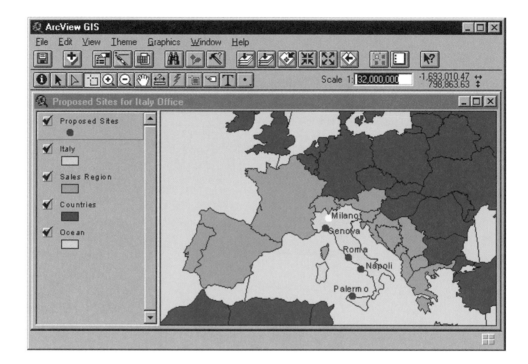

Notice that the view is still centered on Milano, causing part of the Sales Region theme to be cut off. You'll shift the display using the Pan tool.

9. Click on the Pan tool, then move the cursor (now a hand) anywhere over the display. Hold down the mouse button and drag the display (slightly) up and to the left, then release the button. ArcView redraws the view, filling in any blank areas.

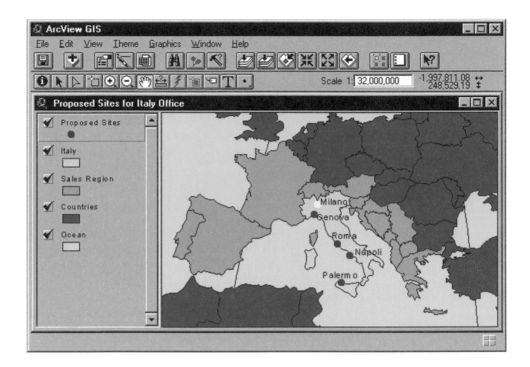

You're ready for your presentation. You've practiced several different ways of zooming in and out and you've set the view scale directly. Your presentation using ArcView's World data and zooming tools will help your audience visualize your sales region and the site you're proposing for the new sales office.

Zooming in and out in a view. ArcView gives you even more ways to zoom in and out. With the Zoom In and Zoom Out *buttons,* you can zoom from the center of the view; with the Zoom In and Zoom Out *tools,* you can zoom from a position or area you define. For more information, search for these Help Topics: *Zooming in and out on a view, Zoom In, Zoom Out, Zoom In tool, Zoom Out tool.*

Setting scale thresholds for themes

With ArcView, you can control the scale at which a theme displays by setting a scale threshold for the theme. When the scale of the view is not within the limits you set, the theme won't display.

Suppose you're a traffic controller for a metropolitan area. You need to display the location of each accident that occurs. In the past, police officers radioed in the location and time of each confirmed accident. This information was then relayed to your office and you manually plotted the accidents on a map.

Your agency has been selected to test a new method of registering accidents as they occur. Each time police officers respond to an accident, they use a GPS (Global Positioning System) to record the location of the accident. (A GPS uses signals sent by satellites to precisely determine a location.) ArcView uses the coordinate information from the GPS to map the accidents as they occur.

Your task is to come up with a way to store all the themes you need in the same view and be able to work with them at different scales. For example, when you examine accidents along city streets, you need to zoom in more than when you examine accidents along major highways. At each scale, you want to display certain themes and not others. To support the different scales you'll be working at, you'll use ArcView's display theme property to set scale thresholds for drawing themes.

Exercise 12b

1. From the File menu, choose Open Exercise. In the Exercises scrolling list, select "ex12b," then click OK. When the project opens, you see a view containing five themes.

ENVIRONMENTAL SYSTEMS RESEARCH INSTITUTE, INC.

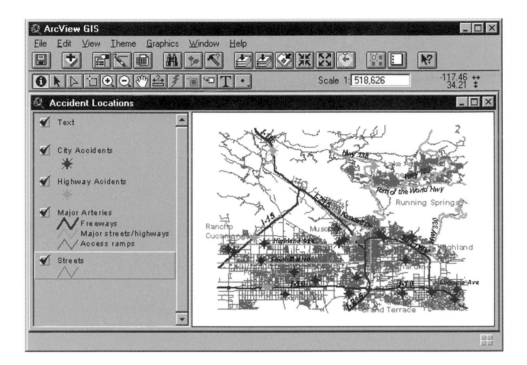

During this exercise, don't maximize or resize the view window, as this alters the view's scale. Your scale values may vary slightly from those reported, depending on your screen's resolution, font size, and other factors.

The Text theme contains the names of towns and major highways; the City Accidents theme contains the locations and times of traffic accidents that occur on city streets; the Highway Accidents theme contains the locations and times of traffic accidents that occur on major highways; the Major Arteries theme contains alternate routes crucial to commuters during rush hour; the Streets theme contains all the city streets for the metropolitan area.

Right away you see that this view is cluttered. At this scale, you can't interpret the information each theme presents without manually turning each theme off and on. To see the highway accidents better, you'll turn some themes off.

2.   Click on the check box for the City Accidents and Streets themes to turn them off.

Now you can easily see where the highway accidents are, which highways they're on, and which cities they're near.

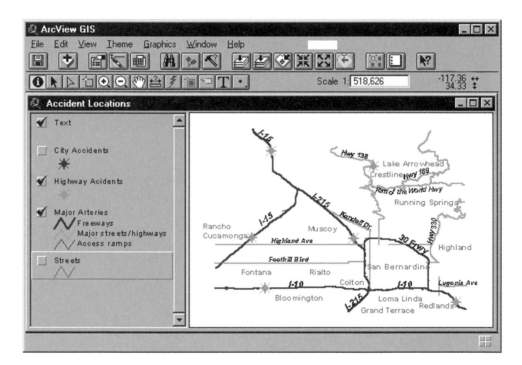

To see the city accidents instead, you'll turn on different themes.

3. Click on the check boxes for the Text, Highway Accidents, and
    Major Arteries themes to turn them off; do the same for the City
    Accidents and Streets themes to turn them on.

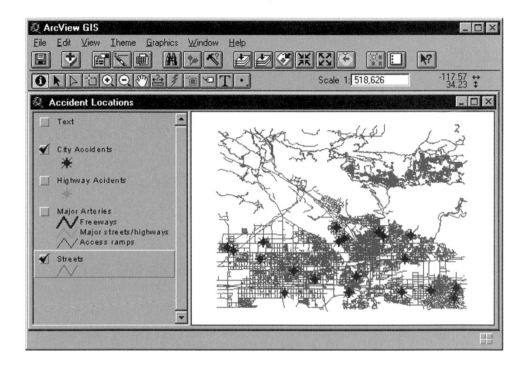

Now you can see the city accidents, but at this scale, you don't clearly
see which streets they're on. To zoom in enough to see which street each
accident is on, you'd have to click the Zoom In button about six times.
Because of the density of the streets, the view takes a long time to draw,
so you won't use this method of zooming in. Instead, you'll set a scale
directly in the Scale box.

4.   Click in the Scale box, drag the cursor to highlight the current value, then change the value to **150,000** and press Enter on your keyboard.

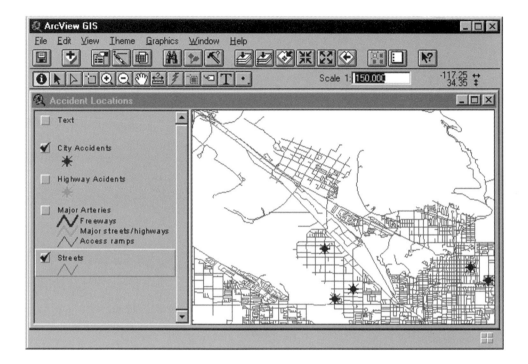

At this scale, you can see the city accidents and the streets they're on. You'll use ArcView's display theme property to set a scale threshold that prevents the streets from drawing until you've zoomed in enough to see them clearly (about 1:150,000 scale).

You'll return to the original view scale.

5. Click on the check boxes for the themes that are turned off to turn them on, then click the Zoom to Full Extent button. ArcView zooms out so you can see all the features in all the themes.

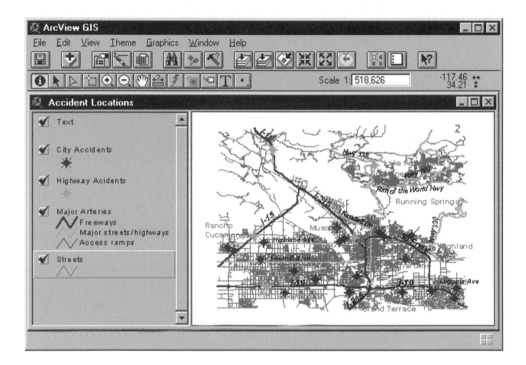

Next you'll set a scale threshold for the Streets theme.

6. With the Streets theme active, select Properties from the Theme menu to display the Theme Properties dialog box.

7. Scroll down to the Display icon along the left margin and click on it. Enter **150000** in the Maximum Scale input field. (Don't enter a value in the Minimum Scale input field.)

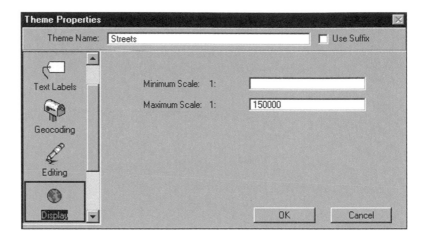

8.   Click OK. Because the value in the Scale box is larger than 150,000, the Streets theme no longer draws in the view, even though it's turned on in the Table of Contents.

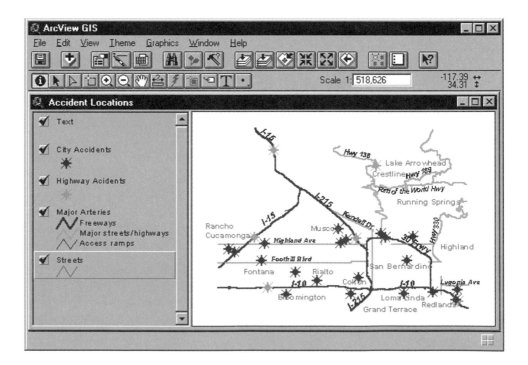

9. Click in the Scale box, drag the cursor to highlight the current value, change the value to **149,999,** then press Enter on your keyboard. When the Scale box contains a value smaller than 150,000, the Streets theme draws.

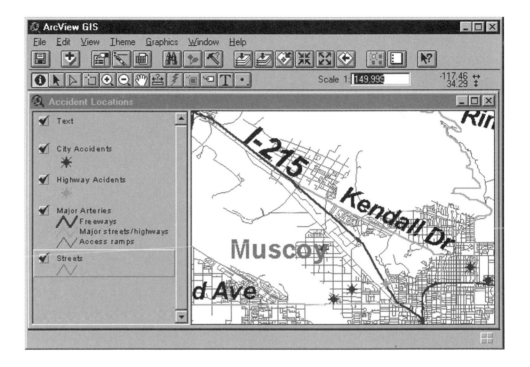

Notice that some of the text labels for the cities and highways cover up accident sites. You don't want them to display when you zoom in to the city streets, so you'll set a scale threshold for the Text theme. First, you'll zoom out again.

10. Click the Zoom to Full Extent button.

11. Make the Text theme active, then select Properties from the Theme menu.

12. Click on the Display icon along the left margin if it's not already selected, then enter **150000** in the Minimum Scale input field.

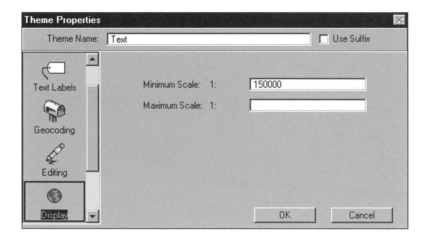

13. Click OK. Now the Text theme won't draw if the value in the scale box is smaller than 150,000.

14. Click in the Scale box, drag the cursor to highlight the current value, enter **149,999,** then press Enter on your keyboard. The Streets theme draws and the Text theme doesn't.

ENVIRONMENTAL SYSTEMS RESEARCH INSTITUTE, INC.

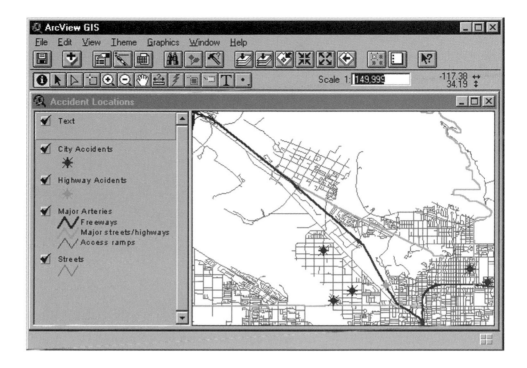

The city accidents are no longer covered up by text and you can see the streets they're on.

Setting maximum and minimum scale thresholds. You can set a range of scales for displaying a theme. For example, if you set a minimum scale threshold of 50,000 and a maximum scale threshold of 100,000, the theme will display only when the value in the Scale box is between these two values. For more information, search for this Help Topic: *Setting a theme's display properties.*

You've completed your task of creating a way to display daily accidents along with other important themes at different scales. Now when you display accidents on major highways, your view won't be cluttered by city streets. Likewise, when you zoom in to see city accidents, they won't be covered up by unnecessary text. By setting scale thresholds for themes, you can display the information you need at the right scale, allowing you to work effectively with a lot of themes in the same view.

If you want to go on to the next chapter, leave ArcView running. Otherwise, choose Exit from the File menu.

SECTION 3

Querying data

The next two chapters introduce you to methods of selecting features and records, and then working with the ones you select. In chapter 13, you'll get information about features, select them directly in a view, then select them according to their attributes, one at a time and in groups. You'll also learn how to select the features you want to show and hide the ones you don't. In chapter 14, you'll select records in a table and work with them by sorting them, promoting them, getting statistics about them, summarizing them, and merging them.

SECTION 3:
Querying data

Selecting map features in a view

Getting information about features

Selecting features based on their attributes

Hiding features

Selecting map features in a view

Sometimes you want to know about a particular feature, or a group of features, rather than about all the features in a theme. With ArcView GIS, you can get information about features by clicking on them one at a time with the Identify tool or by selecting them as a group with the Select Feature tool, then opening their attribute table.

When you want to find features based on their attributes, ArcView lets you find and select them one at a time with the Find button, or as a group using the Query Builder. With Find, you enter an attribute value and ArcView selects the first matching feature it finds. With the Query Builder, you write a statement, called a *query,* specifying an attribute and a value. ArcView selects all the features that match your query.

You can also write a query to define which features in a theme will display in the view. In this case you use the Query Builder in the Theme Properties dialog box to select features. Features that aren't selected won't display in the view.

Getting information about features

In ArcView you can click on a feature in a view with the Identify tool to display its attributes in a dialog box. Identifying features this way is fast. You don't have to select the feature or open the theme's attribute table to see its attributes. But this is only helpful when you want to identify a few features, one at a time. When you want to compare the attributes for a group of features, it's better to select them with the Select Feature tool, then open their attribute table.

When you *select* features, you create a separate set or subset of features. ArcView highlights the selected features in the view and in the theme table. Operations you perform on the theme will affect only the selected set.

Suppose that you're a real estate agent. A family has asked you to show them available properties in your area. They're looking for a tract home, preferably on a corner lot, with three bedrooms.

Exercise 13a

1. If necessary, start ArcView. From the File menu, choose Open Exercise. In the Exercises scrolling list, select "ex13a," then click OK. You see a view with streets, parcels, and a housing tract. The lots in the housing tract are classified into two groups, those that are for sale, and those that aren't.

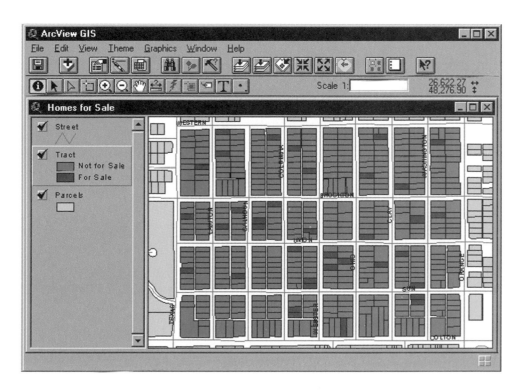

You want to get information about corner lots that are for sale. You'll use the Identify tool.

2.   With the Identify tool selected, click on a corner lot that's for sale. The feature flashes in the view and a dialog box displays.

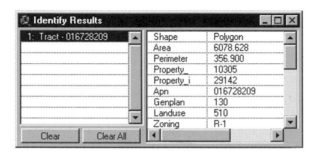

The left side of the Identify Results dialog box lists the feature (lot) you identified and the right side lists its attributes. These are the same attributes that are stored in the theme's attribute table.

3.   Scroll down the list of attributes. The Bd_rms field tells you the number of bedrooms the house has.

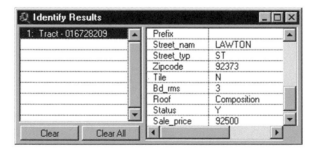

The next to last field in the table, Status, is set to "Y," indicating that the house is for sale. The Tract theme is displayed based on the values in this field.

4.   Click on a few more corner lots that are for sale. Each one is added to the Identify Results dialog box.

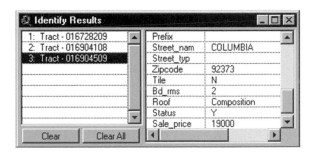

To compare the attributes of all the houses on corner lots, it's more efficient to select them, then open their attribute table. You'll use the Select Feature tool to select them.

5.   Close the Identify Results dialog box, then click on the Select Feature tool. Hold down the Shift key and click on each of the green corner lots in the view (there are eight in all). ArcView highlights the selected lots in yellow.

If you select the wrong feature, you can unselect it by holding down the Shift key and clicking on it again.

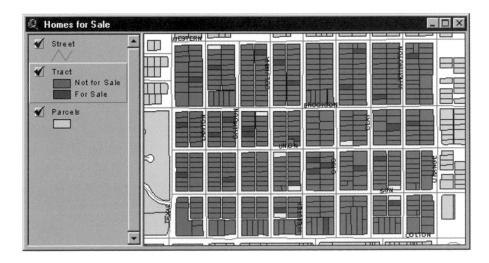

The attributes of the selected lots are also selected and highlighted in the theme's attribute table. To see them, you'll open the table.

6. Click the Open Theme Table button on the View button bar. The Tract theme's attribute table opens.

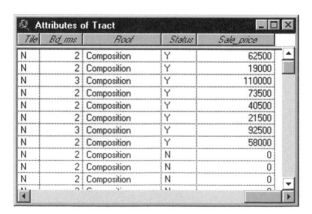

Shape	Area	Perimeter	Property
Polygon	7710.413	381.350	3
Polygon	6839.675	374.619	3
Polygon	6709.897	369.130	3
Polygon	6190.391	361.044	3
Polygon	6673.998	365.817	3
Polygon	6286.618	363.021	3
Polygon	6279.261	361.912	3
Polygon	6342.927	363.245	3
Polygon	6332.976	352.558	3
Polygon	6658.709	361.976	3
Polygon	6475.233	355.410	3

At first, you don't see any highlighted records. To see the entire group of selected records at the same time so you can compare them, you'll use the Promote button to move them to the top of the table.

7. Click the Promote button on the Table button bar. The selected records appear at the top of the table. Now scroll all the way to the right to see the Bd_rms field.

Tile	Bd_rms	Roof	Status	Sale_price
N	2	Composition	Y	62500
N	2	Composition	Y	19000
N	3	Composition	Y	110000
N	2	Composition	Y	73500
N	2	Composition	Y	40500
N	2	Composition	Y	21500
N	3	Composition	Y	92500
N	2	Composition	Y	58000
N	2	Composition	N	0
N	2	Composition	N	0
N	2	Composition	N	0

Now you can see that two of the houses on corner lots have three bed-rooms. To find them in the view, you'll use the Identify tool to click on their records in the table. But first you'll arrange the view and table windows so you can see them both at the same time.

8. Click on the table's title bar, then drag it to the upper left corner of the ArcView window. Now click on the view's title bar, drag it to the lower right corner, and make its window smaller.

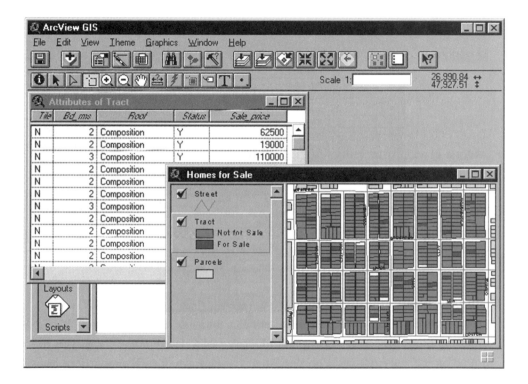

 9. Make the table active, click on the Identify tool in the Table tool bar, then click on the first highlighted record with a value of "3" in the Bd_rms field. The Identify Results dialog box displays and the feature you picked flashes in the view. (If you don't see the feature flash, click on the record again.)

> By default, ArcView places square brackets around field names and quotes around strings. The entire query is enclosed by parentheses.

With this query you could select all the houses that are for sale. Then you could build a second query to select houses from this set that have three bedrooms. Instead, you'll build one query to select houses that are for sale and have three bedrooms, at the same time.

Combining queries. You can combine more than one query by using the And and Or operators to connect them. For example, to select houses that are for sale *and* have three bedrooms, you would use this query: ([Status] = "Y") And ([Bd_rms] = 3). ArcView selects all the houses that meet both criteria. To select all the houses that are for sale *or* have three bedrooms, you would use this query: ([Status] = "Y") Or ([Bd_rms] = 3). In this case, ArcView selects all the houses that meet either criterion. For more information, search for these Help Topics: *Query Builder, Query Builder (Dialog box)*.

---

7. Click the "and" button, double-click "[Bd_rms]" in the Fields list, click the "=" button, then double-click "3" in the Values list. Now your query should look like this:

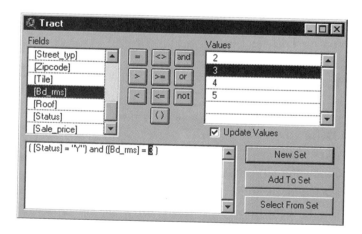

8. Click the New Set button to select those houses that are for sale and have three bedrooms and place them in a new set. ArcView highlights them in the view. (You may need to move the Query Builder dialog box so you can see the view.)

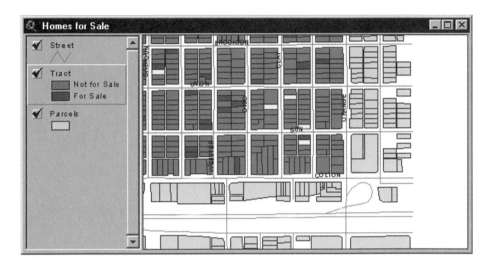

To see all the selected lots, you'll zoom out to them.

 9. With the view active, click the Zoom to Selected button. ArcView zooms out so you can see all the selected features.

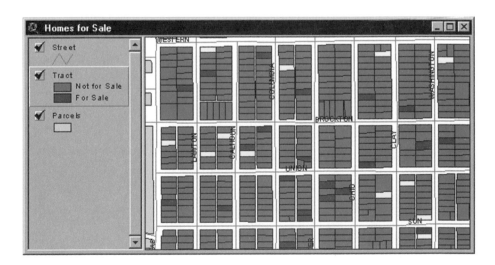

The results are encouraging. There are plenty of three-bedroom houses for sale in the tract to show your clients. Rather than looking at all of them, your clients decide to look at houses priced below $100,000. So, you'll build another query to select houses from the currently selected set, according to the values in the Sale_price field.

10. In the Query Builder's text box, highlight the portion of the query that's between the parentheses (don't highlight the parentheses themselves), then delete it by pressing the Delete or Backspace key on your keyboard.

11. Double-click "[Sale_price]" in the Fields list, click the "<" button, then type **100000.** Your query should look like this:

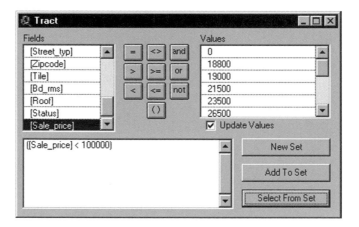

12. Now click the Select From Set button to select houses priced below $100,000 from the currently selected set of three-bedroom houses for sale.

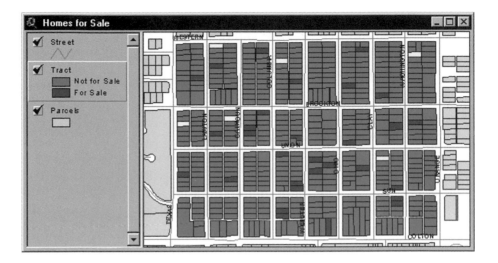

To look at the price of each house, you'll open the theme table.

  13. Close the Query Builder. Click the Open Theme Table button, then the Promote button. ArcView displays the selected records at the top of the table.

14. Scroll all the way to the right to see the price of each house that's selected.

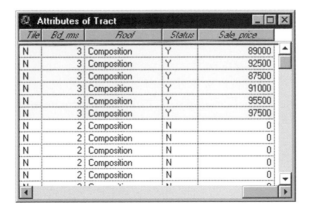

Your clients are satisfied that these houses meet their initial criteria and are ready to look at them.

In this exercise, you used the Find button to locate and select a feature from an attribute value, then the Query Builder to select a group of features by specifying an attribute and the desired value for it. In the next exercise, you'll use the Query Builder from the Theme Properties dialog box to select the features you want to display in the view.

Hiding features

Sometimes you have more features in a theme than you want to work with at a given time. You don't want to delete any data, because you may need it in the future. One way to reduce the size of your data set without deleting any data is through a process called *filtering*. Filtering keeps some of the data visible and hides the rest. To filter data in ArcView, you build a query that selects the features you want to show. Data that's not selected is hidden. When you clear the query, the hidden features return. To build this kind of query, you access the Query Builder from the Theme Properties dialog box, rather than from the View interface.

Filtering helps keep your database manageable. When you filter rather than select data, you reduce the clutter in a view and the time needed to draw it, and make the data easier to interpret and work with. If you have hundreds or even thousands of features in a theme, and you select some of them, it can be hard to distinguish the selected ones from the others. Also, the more records there are in a theme table, the longer it takes ArcView to search for features.

Suppose you're a member of an agricultural commission concerned about the decline of citrus-growing areas due to residential and commercial development. You want to create a map showing the current citrus-growing areas, then use it to track future changes in land use.

Exercise 13c

1. From the File menu, choose Open Exercise. In the Exercises scroll-
    ing list, select "ex13c," then click OK. You see a view showing
    streets and land parcels.

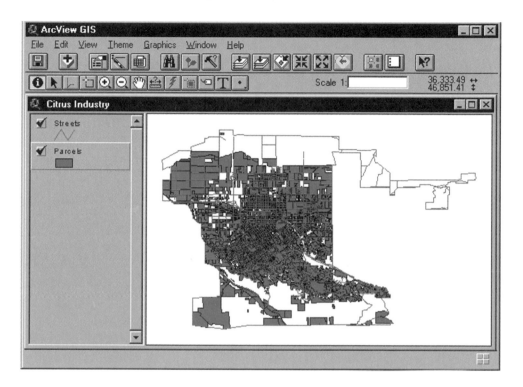

According to the agricultural commission, the land parcels were once
used exclusively for growing citrus and other agricultural products, but in
recent years they've been converted to residential and commercial land
use. Areas where streets are dense (toward the center of the view) have
undergone the heaviest development.

For each feature in the Parcels theme, there is a land use code. A code of
"732" indicates that the land is used for growing citrus. You'll build a
query in the Theme Properties dialog box to filter the Parcels theme so
that only the land parcels used for growing citrus will be displayed.

2. With the Parcels theme active, choose Properties from the Theme menu in the View menu bar. The Theme Properties dialog box displays.

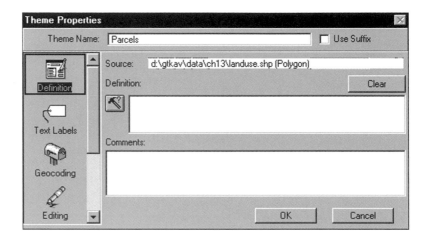

By default, the Definition icon is highlighted on the left. On the right, you see the options for the Definition category of properties.

3. Click the Query Builder button in the Theme Properties dialog box. The Query Builder dialog box displays. It looks like the Query Builder you access from the View interface, but instead of the New Set, Add To Set, and Select From Set buttons, it has an OK button.

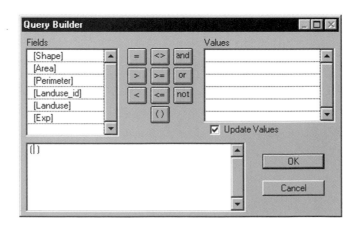

4. In the Fields list, double-click on "[Landuse]." (Because there are a lot of unique values in this field, it takes a while for them to update in the Values list.)

5. Click the "=" button, then scroll down in the Values list until you find "732" and double-click on it. Your query should look like this:

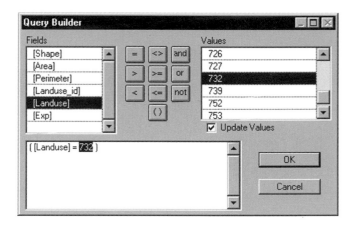

6. Click the OK button in the Query Builder, then click the OK button in the Theme Properties dialog box to select parcels that match your query. ArcView selects, but doesn't highlight, the features that match your query.

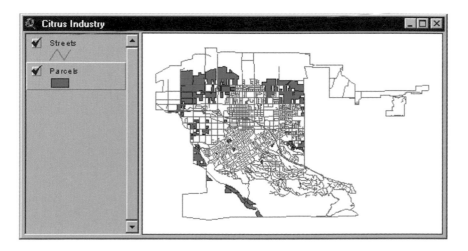

Now the view shows only the parcels that are used for growing citrus; other parcels are hidden from view. You see that the remaining citrus groves are located mainly on the outskirts of the city. Next you'll zoom in to the new Parcels theme.

The area covered by the Parcels theme is defined by the visible features. Features that are hidden in the view are also hidden in the theme table.

 7. Click the Zoom to Active Theme(s) button. ArcView zooms in so the active theme, Parcels, fills the view.

Now you have a theme that shows only the parcels used for citrus. As land use patterns continue to change, you can copy, then update your data to reflect these changes. In this way, you create multiple versions of the data, each one representing a different time period. By filtering each version of the data, you can create a series of maps that show changes in citrus-growing areas over time.

If you want to go on to the next chapter, leave ArcView running. Otherwise, choose Exit from the File menu.

ENVIRONMENTAL SYSTEMS RESEARCH INSTITUTE, INC.

SECTION 3:
Querying data

Selecting records in a table

Selecting records

Summarizing records in a table

Merging features

CHAPTER 14

Selecting records in a table

In the last chapter, you selected features with the Select Feature tool, the Find button, and the Query Builder. In each case, when you selected features, the corresponding records in the theme table were also selected. This is because features and their attributes are linked in ArcView GIS.

The link between features and attributes lets you perform the same operations on either records in a table or features in a view. You can select records directly, search for them with the Find button, or select them with expressions you create in the Query Builder. When you select records, the corresponding features in the view are also selected.

In this chapter, you'll select records with the Select tool and the Query Builder. You'll get statistics on selected records, and *summarize* the theme table to find out how many records share each of the values in a specified field. You'll also learn how to merge several features (and their records) into one.

ENVIRONMENTAL SYSTEMS RESEARCH INSTITUTE, INC.

Selecting records

Suppose you're working as a ranger in the Marsabit district of Kenya. You need to identify drought-threatened areas in your district so that plans can be made to prevent overgrazing by cattle that belong to local farmers. You have a contour map that defines the boundaries of areas that get specific amounts of annual rainfall.

Areas that receive less than 300 millimeters of annual rainfall are considered to be threatened by drought. You'll find these areas by selecting the appropriate records. Next, you'll determine how many square kilometers are affected. Then you'll merge all the threatened areas into a single feature with a single record.

> **Understanding contour maps.** A contour map is a map in which points of equal value (for a specific attribute) are connected by lines. The pattern of the lines shows how the attribute is distributed over the mapped surface. The areas between the contour lines are sometimes shaded, as in the map on the next page, to make a stronger visual impression. Contour maps are commonly used to show the distribution of attributes like temperature, precipitation, atmospheric pressure, and elevation. (Strictly speaking, the term *contour* applies only to maps of elevation; the general term for maps of this kind is *isoline*.)

Exercise 14a

1. If necessary, start ArcView. From the File menu, choose Open
    Exercise. In the Exercises scrolling list, select "ex14a," then click
    OK. When the project opens, you see a map of annual rainfall in the
    Marsabit district.

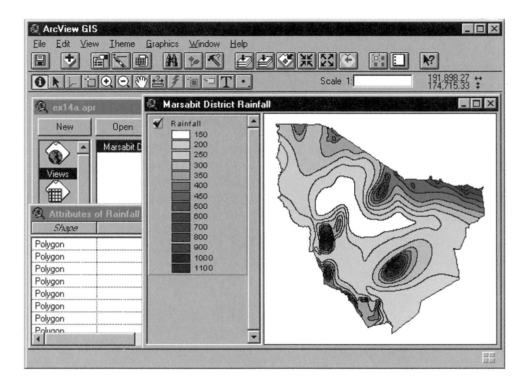

2. Make the Attributes of Rainfall table active and scroll to the right.
    The table contains each polygon's area in square kilometers, the
    amount of rainfall it gets in millimeters, and a description (dry,
    medium, or wet).

It's apparent from the map that much of the land in the district is threatened by drought. Suppose for the moment that you're interested in looking at the attributes of the larger areas.

3. In the Attributes of Rainfall table, scroll back to the left and click on the *Areakm* field to make it active. The cell containing the field name becomes shaded when the field is active.

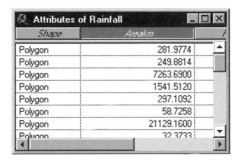

4. Click the Sort Descending button. The records in the table are sorted by area in square kilometers, from largest to smallest.

> **Sorting records.** You can sort records in ascending or descending order of the values in any field. The sort order remains in effect until you close the table, re-sort it, export it, or close the project. For more information, search for this Help Topic: *Sorting a table*.

5. The Select tool should be highlighted by default. (If it isn't, click on it now.) Click anywhere in the first record (which has an Areakm value of 21129.1600) to select it. Hold down the Shift key and click on the next three records. The records are highlighted in the table, and the corresponding features are selected in the view.

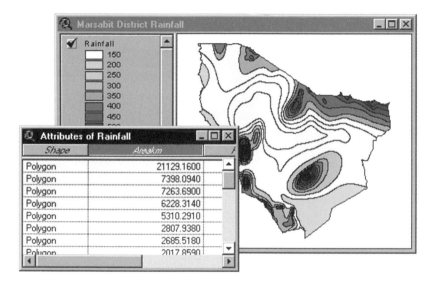

On the left side of the tool bar, ArcView tells you how many records (4 of 67) are selected in the table.

You could now scroll through the table and confirm what your first look at the map suggested: the largest areas also tend to be the driest. What you really need to do, however, is to find all the areas that are susceptible to drought—those that receive less than 300 mm of rainfall annually. You'll do this with the Query Builder. First, you'll clear the currently selected records.

6. Click the Select None button. The highlight disappears from the previously selected records and features.

Selection buttons. To the left of the Select None button is the Select All button, which selects all records in the table. To the right of the Select None button is the Switch Selection button, which unselects all selected records and selects all unselected records.

7. Click the Query Builder button to display the Attributes of Rainfall dialog box. In the Fields scrolling list, double-click [Rainfall] to add it to the query text box. Click the "less than" (<) operator button. Finally, double-click [300] in the Values scrolling list. Your query should look like this:

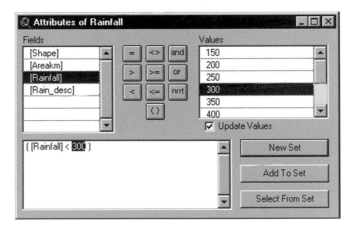

8. Click New Set. All records with rainfall values of less than 300 are selected in the table, and their corresponding features are selected in the view.

9. Click the Promote button to bring all of the selected records to the top of the table.

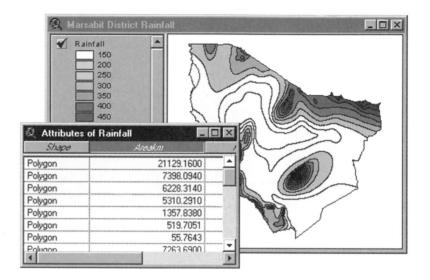

10. Close the Query Builder.

You've selected records for those areas of the district that are subject to drought. You'll work with these selected records in the next exercise.

Summarizing records in a table

You can use a pulldown menu to get statistics at any time about all records or selected records in a table. You can also summarize a table to subtotal records by the values in a specified field. Your subtotals, along with any statistics on them you request, are placed in a new table. You can summarize all records in a table or selected records only.

In the next exercise, you'll get statistics on the total area threatened by drought. Then you'll summarize the attribute table by rainfall value to get a more detailed statistical breakdown.

ENVIRONMENTAL SYSTEMS RESEARCH INSTITUTE, INC.

Exercise 14b

1. From the File menu, choose Open Exercise. In the Exercises scrolling list, select "ex14b," then click OK. When the project opens, you see the Marsabit District Rainfall view and its attribute table. The table is active, and records with rainfall values of less than 300 are selected.

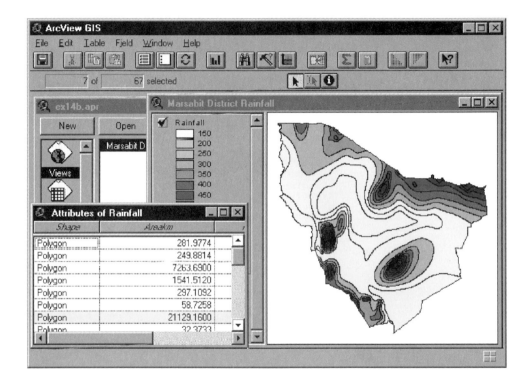

 2. Click the Promote button to move the selected records to the top of the table.

3. In the table, click on the Areakm field to make it active.

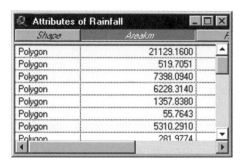

4. Select Statistics from the Field menu. A message box displays statistics on the Areakm field's values. The statistics apply only to the selected records.

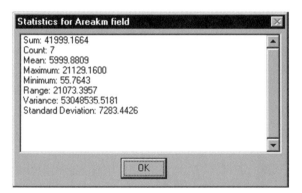

You see that 41,999.1664 square kilometers of the district get less than 300 mm of rainfall annually. You see the largest, smallest, and mean (average) areas of the records in your selected set, as well as other statistical information.

5. Click OK to dismiss the message box.

Now you'll summarize the table to see how the total drought-threatened area is broken down by amount of rainfall received.

Summarizing a table. When you summarize a table, ArcView counts the occurrences of each value in a specified field. It then creates a new table, called a *summary table,* with at least two fields. One of these fields lists the unique values that have just been counted, and the other tells you how many times each unique value was found. Suppose that your attribute table contains ten records, and that a field called *Type* tells you whether each record represents a factory or a farm. When you summarize the table on the Type field, your summary table will contain two records: one with the values "factory" and "3" (assuming there are three factories) and another with the values "farm" and "7." You can also add statistical fields to your summary table. If your attribute table has information on sales and number of employees, a summary table could give you the average number of employees per factory and per farm, or total revenue by farm and factory, or a number of other statistics. For more information, search for these Help Topics: *Summarize (Dialog box), Summarizing a table.*

6. In the Attributes of Rainfall table, scroll to the right and click on the Rainfall field to make it active.

7. From the Field menu, choose Summarize to display the Summary Table Definition dialog box.

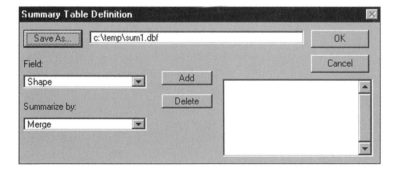

By default, the new table is called "sum1.dbf." (You could change it to something more descriptive if you wanted to.)

If you clicked OK now, your summary table would contain two fields: a Rainfall field and a Count field. Because you want to know how much area is coded with each rainfall value, you will also request statistics on the Areakm field.

8. From the Field drop-down list, select "Areakm." From the Summarize by drop-down list, select "Sum." Click the Add button to place the expression "Sum_Areakm" in the box on the right.

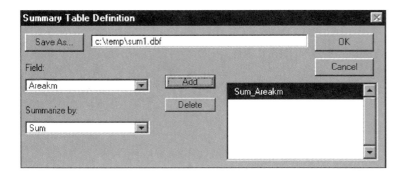

9. Click OK.

ArcView creates a summary table with one record for each unique value (from among the selected records) in the Rainfall field. The summary table has a field of rainfall values, a Count field, and a field that sums the area in square kilometers covered by each rainfall value.

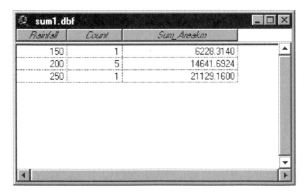

You see that there were three unique rainfall values among the selected records in the attribute table. Two values occurred once and one occurred five times. The Sum_Areakm field tells you how much area gets 150 mm of rainfall annually, how much 200 mm, and how much 250 mm.

10. Close the sum1.dbf table.

In the next exercise you'll merge the seven drought-threatened areas into a single feature.

Merging features

By the same process you use to summarize a table, you can merge features. In a merge, all features with a common value in the active field of the theme table are aggregated into a single feature with a single record. (A single feature may be composed of discontinuous polygons.) ArcView saves the merged features to a shapefile (shapefiles are described in chapters 8, 23, and 24) and displays them as a new theme.

Merging features can make it easier to manage data. (For instance, by merging the state of Maine with its small offshore islands, you could create a theme table with one record instead of perhaps thirty-five.) A merge can also give you a more generalized view of your data, as you'll see in the next exercise.

Exercise 14c

1.   From the File menu, choose Open Exercise. In the Exercises scroll-
     ing list, select "ex14c," then click OK. When the project opens, you
     see the Marsabit District Rainfall view and its attribute table. The
     table is active, and no records are selected.

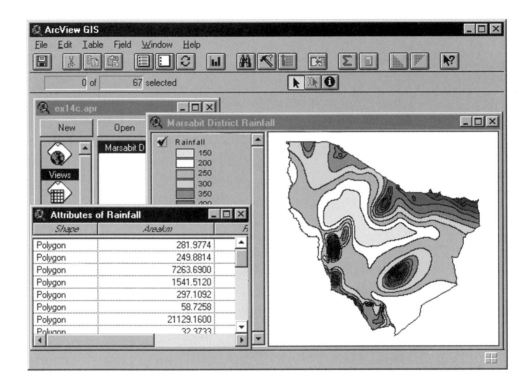

2.   In the table, scroll to the right and click on the Rain_desc field to
     make it active.

The Rain_desc field contains three different values. (Areas receiving less
than 300 mm of rainfall annually are called dry, areas receiving from 300
to 600 mm are medium, and areas receiving 700 mm or more are wet).
The merge operation will create a new theme consisting of three features
with three records in its attribute table. (In this case, you are merging all
records, but you could also merge selected records.)

ENVIRONMENTAL SYSTEMS RESEARCH INSTITUTE, INC.

3. Click the Summarize button to display the Summary Table Definition dialog box.

4. The default selections in the Field and Summarize by drop-down lists are "Shape" and "Merge," respectively. (If these are not the selections, select them now.) Click the Add button to place the expression "Merge_Shape" in the Summary statistics box on the right.

This is all you need to do to merge features. As with any summarize operation, however, you can also request statistics.

5. From the Field drop-down list, select "Areakm." From the Summarize by drop-down list, select "Sum." Click the Add button to add the expression "Sum_Areakm" to the box on the right.

6. Click the Save As button to navigate to the *drive:\directory* where you want to save the new table (and shapefile) ArcView will create. Rename the table "rain_mrg.dbf," then click OK.

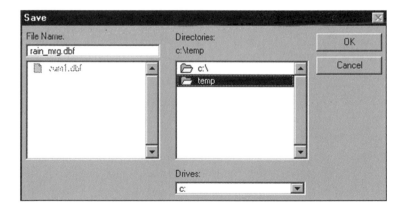

Your Summary Table Definition dialog box should resemble the one in the next graphic (you may be saving to a different directory).

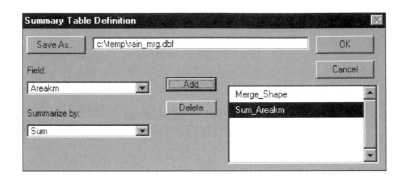

7. Click OK in the Summary Table Definition dialog box to merge features.

ArcView creates a new theme and asks you whether you want to add it to the current view or to a new view.

8. Click the <New View> option to highlight it.

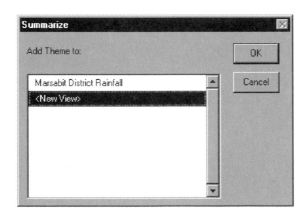

9. Click OK.

ArcView creates a new view with a new theme, Rain_mrg.shp.

10. Click on the new theme to make it active, then click on its check box to turn it on.

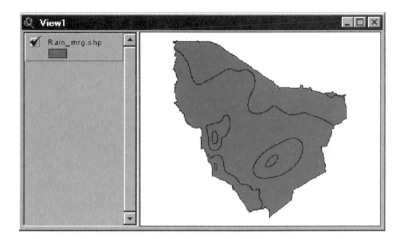

The new theme contains just three features: one feature representing all dry areas, one all medium areas, and one all wet areas. This will be easier to see if you classify your map.

11. Double-click on the theme to open the Legend Editor. From the Legend Type drop-down list, choose "Unique Value." From the Values Field drop-down list, select "Rain_desc" as the field to classify on.

You'll change the default colors to more meaningful colors.

12. Double-click on the symbol representing medium areas to open the Fill Palette in the Symbol Window. In the Fill Palette, click the Color button to change to the Color Palette. Click on a green color square to select it.

13. Click on the Legend Editor to make it active, and double-click on the symbol representing wet areas. Change it to a deep blue color in the Color Palette, then close the Color Palette.

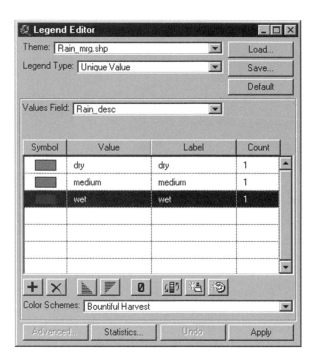

14. Click Apply to apply your changes to the view, then close the Legend Editor and the Symbol Window.

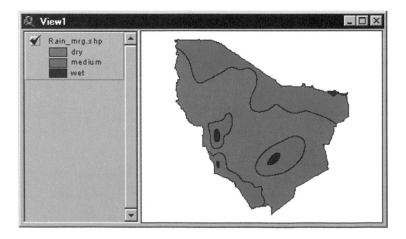

The new map is less detailed than the original contour map, but it makes it easier to see the drought-threatened area's size. Although the medium and wet areas on the map are discontinuous, they are nevertheless single features, as you will see.

15. From the Window menu, select Attributes of Rain_mrg.shp to display the theme table. Resize the table so you can see all the fields.

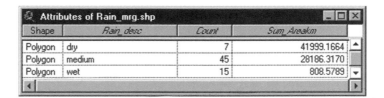

The table contains three records, one for each unique value in the Rain_desc field of the Attributes of Rainfall table (Rain_desc is the field you made active in step 2). The Count field tells you how many records from that table had each value. The Sum_Areakm field (which you requested in step 5) gives you the area in square kilometers coded with each rainfall value.

16. Make sure the Select tool is active and click on the "medium" record.

The record is highlighted in the table, and the corresponding area is highlighted on the view (move the table out of the way if necessary). You can see that ArcView regards the separate polygons as a single feature.

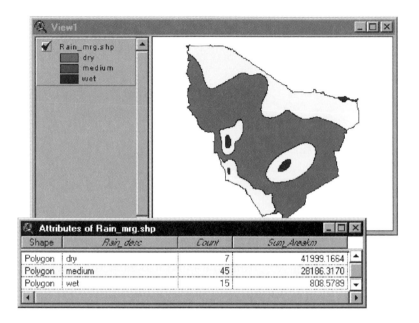

17. Click on the Marsabit District Rainfall view to make it active. Note that the feature merge has not affected your original theme. It has simply given you another way to look at your data.

Now you know which areas in your district are vulnerable to drought, you know the size of these areas, and you have a view that displays them as a single feature. With this information, you can make informed decisions about where and how to restrict cattle grazing.

If you want to go on to the next chapter, leave ArcView running. Otherwise, choose Exit from the File menu.

SECTION 4

Managing tabular data

The next two chapters introduce you to working with tabular data in ArcView GIS. In chapter 15, you'll learn how to hide fields, assign alias names to fields, add fields, and calculate the values for a new field based on existing fields. You'll also learn how to get statistics for a field and summarize a table based on the values in a particular field. In chapter 16, you'll learn how to add tabular data to an ArcView project, then join or link the data to the theme table, depending on the relationship between records. You'll also learn how to access additional data by setting up hot links.

SECTION 4:
Managing tabular data

Displaying and editing tables

Changing the table display

Editing values and adding fields

Summarizing a table

Displaying and editing tables

In chapter 14, you selected and sorted records in a table, viewed statistics, and summarized selected records. In this chapter, you'll modify the appearance of a table by hiding fields, assigning alias field names, and changing field width display. You'll also edit values in a table, add a new field, and use the Field Calculator in ArcView GIS to assign values to the new field. Finally, you'll review the table summary process by summarizing all records in the theme table.

Changing the table display

Suppose that you work for a firm that analyzes consumer behavior to help its clients reach specific markets. You've been hired by a luxury car dealer who is thinking of opening a new dealership in Clayton County, Georgia. Your job is to determine the size of the potential market and to break it down by the townships within the county.

You'll use a set of commercial data in which U.S. households are divided into consumer profiles called *segments*. Fifty segments are distinguished on the basis of demographic and lifestyle factors, such as income, age, education, and number of vehicles per household. Ten of these segments (those that passed your initial cut) are stored as fields in a theme attribute table. Because you've decided that only two of these ten ultimately meet your criteria, you want to modify the appearance of the table.

Exercise 15a

1. If necessary, start ArcView. From the File menu, choose Open Exercise. In the Exercises scrolling list, select "ex15a," then click OK. When the project opens, you see a view with one active theme, Clayton County. The theme displays census tract boundaries for Clayton County, Georgia, and is classified by the townships within the county.

ENVIRONMENTAL SYSTEMS RESEARCH INSTITUTE, INC.

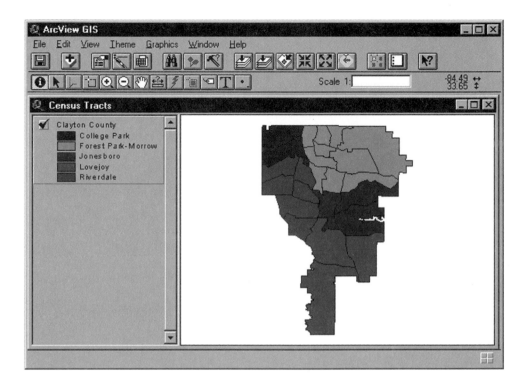

 2. Click the Open Theme Table button to open the Attributes of Clayton County theme table.

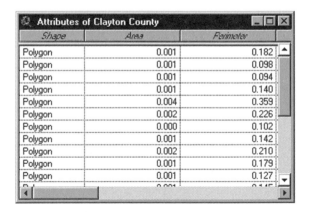

3. Scroll to the right to see the contents of the table.

The table contains one record for every census tract in Clayton County. There are shape, area, perimeter, and ID fields, a field of census tract numbers, and a field listing the township in which each census tract lies.

The table also contains ten fields of consumer profile segments, Hh_seg1 to Hh_seg10. Two of the fields, Hh_seg2 and Hh_seg3, represent consumer households that you've concluded are most likely to buy luxury cars.

You'll hide the eight market segment fields that won't be included in your analysis, as well as most of the other fields.

4.   Make sure the table is active. From the Table menu, select Properties to display the Table Properties dialog box.

The name of the table is displayed in the Title box. (You could change it by typing a new name.) The name of each field appears in a scrolling box at the bottom of the dialog box. In the Visible column to the left of the field name, a check mark indicates whether the field is displayed or hidden. All fields are presently displayed.

5.   Click on the appropriate check marks to hide the following fields: Area, Perimeter, Clayton_id, and Hh_seg1. The check marks disap-

pear, indicating that these fields won't be displayed in the table. Scroll down and click on the check marks next to Hh_seg4 through Hh_seg10 to hide these fields as well. (To make a field visible again, just click once more in the same cell.)

When you hide a field in a table, you may be unable to carry out operations that use the information in that field. For example, if you hide the Shape field, you can't merge features (chapter 14) or convert a theme to a shapefile (chapter 23).

To the right of the Field names is an Alias column. An alias is a descriptive name given to a field that is named with a code or an abbreviation. Your household segment data comes with a data dictionary (a reference document that explains the data in detail), and you'll borrow aliases from its descriptions of the segments.

6. Scroll up toward the top of the box. Click in the Alias cell next to Hh_seg2 and type **Lap of Luxury.** Click in the Alias cell next to Hh_seg3 and type **Established Wealth.**

7. Click OK to apply your changes to the theme table. Widen the table so you can see all fields.

Shape	Tract	Township	Lap of Luxury	stablished Wea
Polygon	13063040100	College Park	0	0
Polygon	13063040301	Forest Park-Morrow	0	0
Polygon	13063040302	Forest Park-Morrow	2	0
Polygon	13063040303	Forest Park-Morrow	0	0
Polygon	13063040402	Forest Park-Morrow	0	0
Polygon	13063040200	College Park	0	0
Polygon	13063040304	Forest Park-Morrow	0	0
Polygon	13063040305	Forest Park-Morrow	0	0
Polygon	13063040401	Forest Park-Morrow	0	0
Polygon	13063040504	Riverdale	0	0
Polygon	13063040503	Riverdale	0	0

Attributes of Clayton County

The hidden fields are no longer displayed, and the names "Hh_seg2" and "Hh_seg3" are replaced by their aliases, "Lap of Luxury" and "Established Wealth."

The aliases are a little too long, so you need to widen the display width of those two fields.

Changing a field's display width doesn't change the amount of data it can hold. The amount of data that can be stored is determined by the character width of a field. This width is already set for the fields in tables that you bring into ArcView. When you add a new field to a table, you specify the character width you want. Once set, the character width of a field can't be changed.

---

8. Position the cursor over the vertical line separating the Lap of Luxury and Established Wealth field names. The cursor changes to a double-headed arrow. Click and drag the cursor to the right to widen the field's display width.

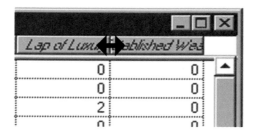

When you resize a field, the new display size remains in effect as long as the project is open and is saved when you save the project.

---

9.   Move the cursor to the right-hand edge of the Established Wealth field and widen it in the same way. (You may need to enlarge the table slightly to make room for the new field display widths.)

Your table looks like this:

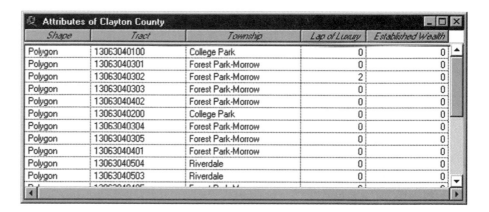

Shape	Tract	Township	Lap of Luxury	Established Wealth
Polygon	13063040100	College Park	0	0
Polygon	13063040301	Forest Park-Morrow	0	0
Polygon	13063040302	Forest Park-Morrow	2	0
Polygon	13063040303	Forest Park-Morrow	0	0
Polygon	13063040402	Forest Park-Morrow	0	0
Polygon	13063040200	College Park	0	0
Polygon	13063040304	Forest Park-Morrow	0	0
Polygon	13063040305	Forest Park-Morrow	0	0
Polygon	13063040401	Forest Park-Morrow	0	0
Polygon	13063040504	Riverdale	0	0
Polygon	13063040503	Riverdale	0	0

Changes to a table's appearance are specific to the project they're made in. If you add the Clayton County theme to another project, the theme table will look as it did at the beginning of this exercise.

By hiding fields, assigning aliases, and widening field displays, you've changed the appearance of the table to suit your needs. In the next exercise, you'll edit values and add a new field to the table.

Editing values and adding fields

You can edit ArcView tables in various ways. You can change cell values and add and delete both records and fields. You can also perform mathematical, logical, and text operations on existing fields and store the results in a new field.

In this exercise, you'll change an incorrect data value. Then you'll add to the table a new field containing the sum of the households in the Lap of Luxury and Established Wealth fields. This number will tell you how many potential luxury car buyers live in Clayton County. (In the next exercise, you'll find out how this market is distributed by township.)

Exercise 15b

1. From the File menu, choose Open Exercise. In the Exercises scrolling list, select "ex15b," then click OK. The project looks as it did at the end of the previous exercise.

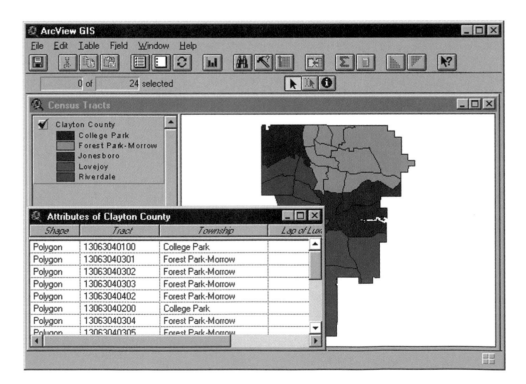

ENVIRONMENTAL SYSTEMS RESEARCH INSTITUTE, INC.

Suppose you've been told that one of the census tracts in your map belongs to Jonesboro, not Lovejoy. As it happens, there are a number of luxury car buyers in this tract, so the error is relevant to your analysis (in any case, you want your map to be accurate).

To correct the mistake, you'll edit the township value of the appropriate record in the table. Before you edit tabular data (change cell values or add or delete records or fields), ArcView requires that you turn on table editing.

2. Make sure the theme table is active. From the Table menu, select Start Editing. (The field names become non-italic, confirming that the table is editable.)

3. Scroll to the bottom of the table. The third record from the bottom (with the census tract number 13063040607) is the one you're going to change.

4. Click on the Edit tool. As you move the cursor over the table, it changes to a pointing hand.

5. Click on the "Lovejoy" value in the Township field of the third record from the bottom. The value is highlighted in black. Type the name **Jonesboro.**

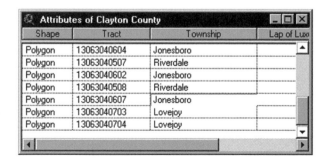

6. Press the Enter key to commit your edit. The cursor moves down to the next cell.

7. From the Table menu, select Stop Editing. The Stop Editing dialog
    box asks you if you want to save your edits.

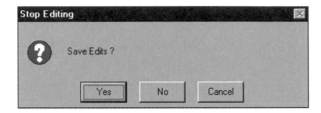

8. Click Yes to save your edits.

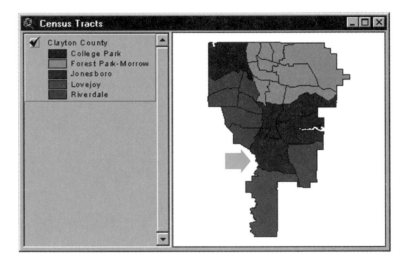

Look at the view. The census tract whose value you edited (shown by an
arrow) is now blue (Jonesboro) instead of green (Lovejoy). When you
saved edits, ArcView reclassified the view.

Now you want to add a new field to the table and populate it with values
that sum the Lap of Luxury and Established Wealth fields for each
record.

9.   With the table active, from the Table menu, select Start Editing.

(In step 7, you stopped editing in order to see the results of your edits. You could have gone on to make all your table edits in a single editing session.)

10. From the Edit menu, choose Add Field to display the Field Definition dialog box.

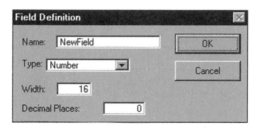

11. In the Name input box, highlight the default name, "NewField," and type **Car_buyers.** Leave the other settings unchanged.

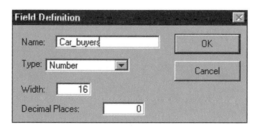

12. Click OK to add the new field to the table.

Defining fields. You can define four different types of fields in ArcView. In a *numeric* field, you can specify the field width (the number of places the field will accept) and the number of decimal places. In a *string* field, you can specify the width. A *Boolean* field accepts values of true or false. A *date* field accepts dates in YYYYMMDD format. For more information, search for this Help Topic: *Field Definition (Dialog box).*

The new field is added at the end of the table and is active.

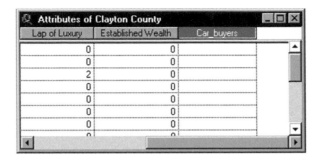

You want the Car_buyers field to contain the sum of the households in the Lap of Luxury and Established Wealth fields. Then you'll know the total number of potential luxury car buyers in each census tract. You can use the Field Calculator to have ArcView do this for you.

 13. Click the Calculate button to open the Field Calculator dialog box. (The Calculate button is available only when table editing is turned on and a field is highlighted in the table.)

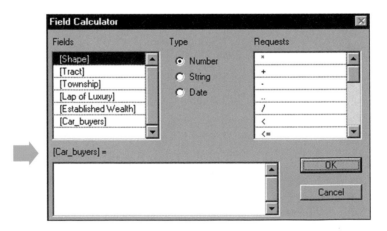

The Field Calculator looks somewhat like the Query Builder, which you used in chapters 13 and 14. With the Field Calculator, you create an expression that tells ArcView how to assign values to the active field. The first part of the expression, [Car_buyers] =, is supplied by ArcView. You'll complete the expression in the expression box.

14. In the Fields scrolling list, double-click [Lap of Luxury] to add it to the expression box. In the Requests scrolling list, double-click the plus sign (+). In the Fields scrolling list, double-click [Established Wealth].

Your completed expression is [Car_buyers] = [Lap of Luxury] + [Established Wealth]. This expression will sum, for each record, the values in the Lap of Luxury and Established Wealth fields and place the results in the Car_buyers field.

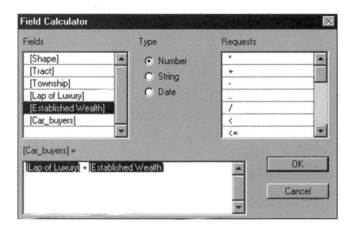

15. Click OK to calculate the values (the Field Calculator dialog box disappears). Then scroll down through the table to see that ArcView has added the Lap of Luxury and Established Wealth values in the Car_buyers field.

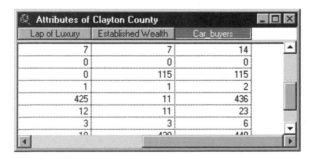

Understanding the Field Calculator. With the Field Calculator you can create expressions ranging from the simple to the complex. The simplest type of expression assigns a single value to every cell in the active field (or, if there is a selected set of records, to the cells in the selected set). For instance, the value "100" could be assigned to each cell in a numeric field, or the value "reads computer magazines" to each cell in a string field. More complex expressions use arithmetic and logical operators to modify values found in other fields of the table. For example, you can add, subtract, multiply, or divide numbers in one field by those in another or by constants. You can search and replace text strings and change the case of text. Over a hundred analytical operations are supported. For more information, search for these Help Topics: *Field Calculator (Dialog box), Calculating a field's values, Performing operations on number fields, Performing operations on string fields, Performing operations on boolean fields, Performing operations on date fields.*

16. From the Table menu, choose Stop Editing. The Stop Editing dialog box asks if you want to save your edits.

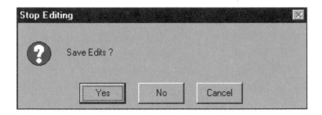

ENVIRONMENTAL SYSTEMS RESEARCH INSTITUTE, INC.

17. Click Yes to save your edits.

The Car_buyers field is now more useful to you than the Lap of Luxury and Established Wealth fields. You could hide these two fields in the table, but instead you will move the Car_buyers field to the left, so that you don't have to scroll so far to see it.

18. The Car_buyers field should still be active. Click on the title bar of the Car_buyers field to make it inactive (not highlighted). Now click and drag the title bar to the left. When its outline is positioned over the Lap of Luxury field, release the mouse button. Scroll to the left to see that the Car_buyers field is placed between the Township and Lap of Luxury fields. (If it isn't, try again.)

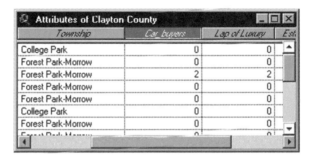

You can change the content of a table by editing it. You can change the way the table displays by hiding fields, assigning aliases, and rearranging the order of the fields. Changes to the table display appear only in the current project, but edits to the table will appear in any project that contains the table.

Summarizing a table

The Car_buyers field represents those households that are most likely to buy luxury cars. You'll use ArcView's Summarize function (which you used on a selected set of records in chapter 14) to see how these households are distributed among the townships of Clayton County.

Exercise 15c

1. From the File menu, choose Open Exercise. In the Exercises scrolling list, select "ex15c," then click OK. The project looks as it did at the end of the previous exercise, except that the display width of some of the fields has been narrowed, and the Car_buyers field is not active.

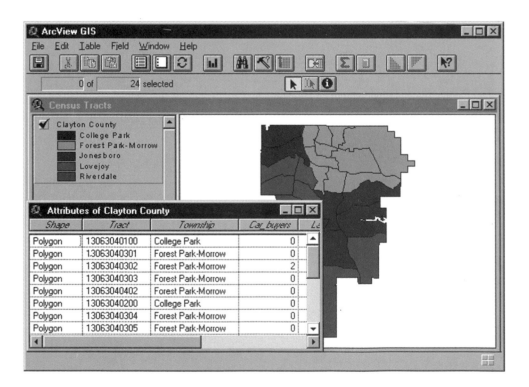

The Car_buyers field breaks down the potential luxury car market by census tract, but your client is more interested in knowing which township to locate his business in. By summarizing the table on the township field, you can find out how the car-buying households are distributed among the five townships of Clayton County.

2.   In the Attributes of Clayton County table, click on the title bar of the Township field to make it active.

3.   Click the Summarize button to open the Summary Table Definition dialog box.

4.   From the Field drop-down list, select "Car_buyers." From the Summarize by drop-down list, select "Sum." Click the Add button to place the expression "Sum_Car_buyers" in the box on the right.

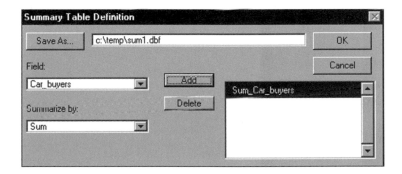

ArcView will create a new summary table with one record for each unique value in the Township field. The summary table will contain a field of Township values and a Count field. (Each township will be counted once for every census tract it contains.) The third field in the summary table (Sum_Car_buyers) will give you the number of potential luxury car-buying households in each township.

5.   Click OK to create the summary table.

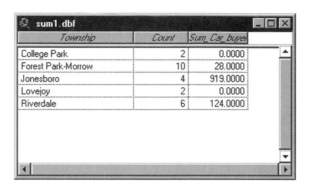

You can see that the township of Jonesboro contains by far the greatest number of potential luxury car buyers. Jonesboro is probably the best location for a new dealership.

You might want to present the results of your analysis in a map. To display the distribution of potential buyers by township, you would merge features (see exercise 14c) on the township field of the Attributes of Clayton County table. The merge would create a new theme of township boundaries. You would then *join* sum1.dbf to the attribute table of your new township theme. (You'll learn how to join tables in the next chapter.) After joining the tables, you would use the Legend Editor to classify the townships by graduated color on the Sum_Car_buyers field.

If you want to go on to the next chapter, leave ArcView running. Otherwise, choose Exit from the File menu.

SECTION 4:
Managing tabular data

Joining, linking,
and hot linking

Joining tables

Linking tables

Creating hot links

CHAPTER 16

Joining, linking, and hot linking

In chapter 8, you learned that every record in a theme table represents a feature in a view. The attributes of each record specify the locations of the features and contain other descriptive information. It's rarely the case, however, that a theme table has all the attributes you might ever want to use to classify or query the features in a view. Often, you'll have information stored in tables that don't belong to any particular theme. To associate these tables with features in a view, you *join* or *link* them to a theme table.

Both joining and linking require that ArcView GIS match records between the two tables. For this to be possible, the tables must share a field of common information. Matches are made between records that have the same attribute value in their common field. In a join, the attributes of records in one table (called the *source* table) are appended to matching records in the other table (called the *destination* table). In a link, the selection of a record in the destination table results in the selection of matching records in the source table.

The *hot link* function in ArcView allows you to link a feature in a theme to a text file, an image, an ArcView document (a view, table, chart, or layout), or another ArcView project. Clicking on the feature then displays the data source to which it's hot linked.

Joining tables

When two tables are joined, they're displayed as one. If the destination table in your join is a theme table, you can query and classify features in the view on any attribute from either original table. A join is therefore a very powerful tool, but it isn't appropriate in all circumstances. To understand why, you need to know something about table relationships.

There are three different kinds of relationships between tables. In a *one-to-one* relationship, a single record in the destination table matches, or corresponds to, a single unique record in the source table. (A destination table of states and a source table of state capitals have a one-to-one relationship.) In a *many-to-one* relationship, several records in the destination table correspond to a single record in the source table. (A destination table of counties and a source table of states have a many-to-one relationship.) In a *one-to-many* relationship, one record in the destination table corresponds to several records in the source table. (A destination table of buildings and a source table of tenants in the buildings have a one-to-many relationship.)

In a join operation, each record in the destination table receives the attributes of its matching record in the source table. If there are many matching records in the source table, this presents a problem, because the attributes of only one of them can be appended. For this reason, tables are usually joined only when the relationship between them is one-to-one or many-to-one. ArcView will perform a one-to-many join if you ask it to, but the join won't preserve the integrity of the one-to-many table relationship. A better option, when table relationships are one-to-many, is to link the records. You'll find out more about linking in exercise 16b.

Suppose you're a staff writer for a monthly magazine about California. You've just been assigned to write a feature article on the best counties to live in. You hope this means lots of travel and visits to some choice spots, like Marin. But your editor wants county-by-county rankings, based on hard data and geographic analysis. It looks like you'll be spending the next few weeks at your desk—but at least you've got ArcView GIS.

Exercise 16a

1. If necessary, start ArcView. From the File menu, choose Open Exercise. In the Exercises scrolling list, select "ex16a," then click OK. When the project opens, you see a view of the state of California with county boundaries.

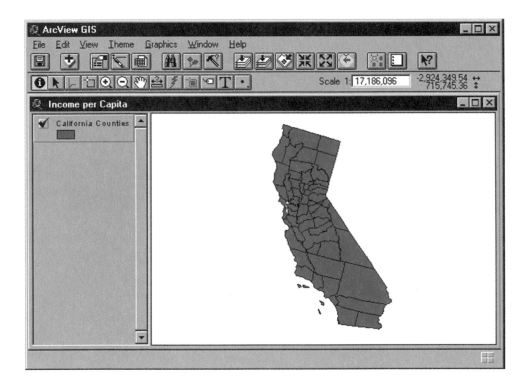

To begin your analysis, you'll evaluate two criteria, economic status and recreational opportunities. Your data includes a table of per capita income values (by county) and a table containing the names of recreation areas (such as national parks) in each county.

Your first goal is to create a map classifying the counties by per capita income.

ENVIRONMENTAL SYSTEMS RESEARCH INSTITUTE, INC.

2. Click the Open Theme Table button to display the Attributes of California Counties theme table.

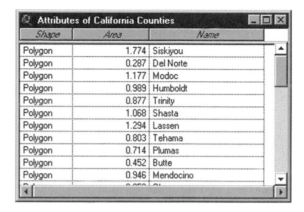

The theme table doesn't contain an income attribute. You'll add a table with income information to the project, then join it to the theme table.

3. Make the Project window active by selecting *ex16a.apr* from the Window menu. Click on the Tables icon if it's not already highlighted. Click the Add button to display the Add Table dialog box.

4. In the dialog box, select the drive where you installed the data for this book, then navigate to *\gtkav\data\ch16* in the Directories list. Click on the "income.dbf" file to select it.

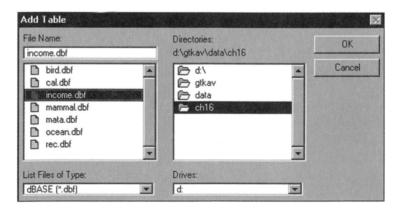

5.   Click OK to add the income.dbf table to your project.

Fips	Cnty_name	Inc_p_cap
6015	Del Norte	10625
6093	Siskiyou	11610
6023	Humboldt	12436
6049	Modoc	10971
6105	Trinity	10781
6089	Shasta	12381
6035	Lassen	12626
6103	Tehama	10990
6045	Mendocino	12776
6063	Plumas	12952
6007	Butte	12083

When the income.dbf table opens, you see that the last field contains per capita income values for each county. The Cnty_name field contains the same data as the Name field in the theme table. You'll use this common field to join the two tables.

> **Understanding the common field.** A common field doesn't have to have the same name in both tables. In fact, the only requirement ArcView imposes is that the field types be the same—for instance, that both fields be string fields or that both be numeric. Since records are matched, however, only when they have the same attribute value in the common field, joining tables on a "common" field of unrelated data won't get you very far. In any join operation, ArcView ignores source table records that have no matching records in the destination table. Destination table records with no matching source table record have blank attributes appended to them. For more information, search for these Help Topics: *Join, Joining tables.*

Note that there is one record per county in both the theme table (which will be the destination table) and the income.dbf table (which will be the source table). Because the relationship between the tables is one-to-one, it's appropriate to join them.

To see both tables during the join, you'll reposition them.

6. Move the income.dbf table to the upper left corner of the ArcView window. Move the Attributes of California Counties table to the lower right corner of the ArcView window.

Now you'll select the common field.

7. Make the income.dbf table active. Click on the Cnty_name field title to make it active.

8. Make the theme table (Attributes of California Counties) active. Click on the Name field title to make it active.

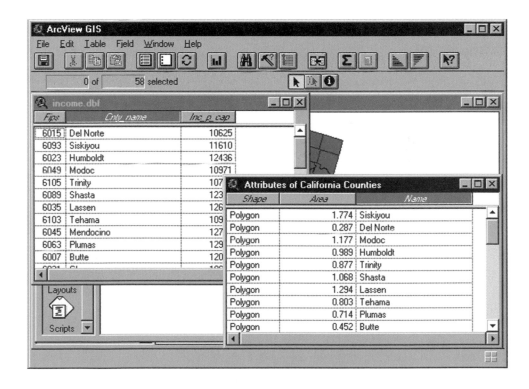

In a join, the active table is always the destination table. The inactive table is the source table. (If there is more than one inactive table, the source table is the one that was most recently active.)

9. With the theme table active, click the Join button. ArcView appends the attributes of records in the income.dbf table to the theme table and closes the income.dbf table.

10. Enlarge the theme table window. Notice that it now contains the fields and attribute values from the income.dbf table.

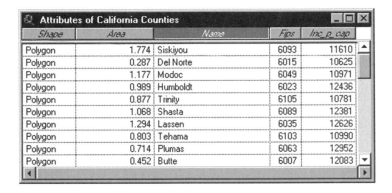

Shape	Area	Name	Fips	Inc_p_cap
Polygon	1.774	Siskiyou	6093	11610
Polygon	0.287	Del Norte	6015	10625
Polygon	1.177	Modoc	6049	10971
Polygon	0.989	Humboldt	6023	12436
Polygon	0.877	Trinity	6105	10781
Polygon	1.068	Shasta	6089	12381
Polygon	1.294	Lassen	6035	12626
Polygon	0.803	Tehama	6103	10990
Polygon	0.714	Plumas	6063	12952
Polygon	0.452	Butte	6007	12083

Now you can classify the counties using the joined data.

11. Close the theme table. Make the view active, then double-click on the theme name in the Table of Contents to open the Legend Editor. (You may need to move the Legend Editor so it doesn't obscure the view.)

12. In the Legend Editor, select "Graduated Color" from the Legend Type drop-down list and "Inc_p_cap" from the Classification Field drop-down list. Select "Green monochromatic" from the Color Ramps drop-down list.

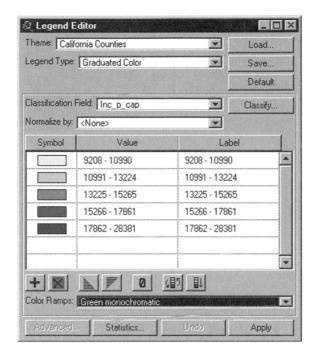

ArcView groups the values in this field into five classes using the Natural Breaks method (see chapter 9) and assigns a symbol to each class using the green monochromatic color ramp.

13. Click Apply in the Legend Editor to apply your changes to the view. Close the Legend Editor.

The view shows the counties of California classified by per capita income, an attribute that wasn't part of the original theme table. Neither of the original tables, however, is changed in any way; the joined table is *virtual*, meaning that it can be displayed and queried, but isn't saved as a separate file. The source table, which was closed during the join operation, remains in the project. You can choose Remove All Joins from the Table menu to restore the theme table to its original appearance.

Joining and linking multiple tables. You can continue joining tables, one at a time, to the same destination table, as long as none of the source tables contains a join. To use a joined table as a source table, first export it by choosing Export from the File menu, then add the exported table to the project. (Alternatively, you can remove the join in the source table, then join the unjoined tables to your destination table.) Multiple tables can be linked in the same way. For more information, search for these Help Topics: *Join, Joining tables, Link, Linking tables, Exporting a table, Export Table (Dialog box)*.

Now that you've classified the counties by per capita income, you want to identify the number of recreational areas in each county. The best places to live should have an above-average number of recreational areas.

If you want to go on to the next exercise, leave the project open.

Linking tables

In the last exercise, you used Join to relate each county to an income value. Now you want to relate each county to the recreation areas it contains. You'll add another table, containing recreation areas by county, to the project. Because most of the counties have more than one recreation area, there's a *one-to-many* relationship between the theme table (destination) and the recreation table (source). When a one-to-many relationship exists between tables, you normally link rather than join them.

Exercise 16b

1. If *ex16a.apr* is still open, make the Recreational Resources view active by selecting it from the Window menu. Otherwise, choose Open Exercise from the File menu, select "ex16b," and click OK. When the project opens, you see the Recreational Resources view.

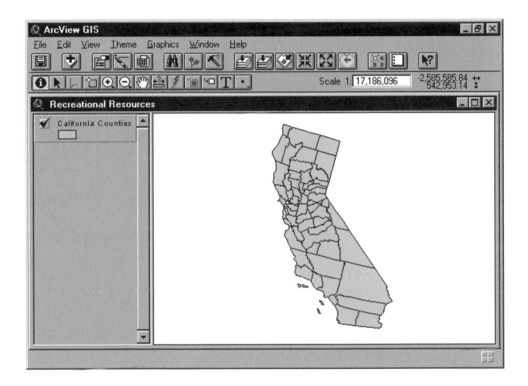

    2. Click the Open Theme Table button to open the theme table.

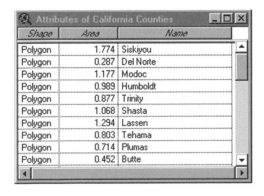

Although this theme table has the same name and contains the same data as the theme table in the last exercise, it's actually a different table. The California Counties theme has been added to the project twice (once in each of the two views). When a theme is added twice, ArcView treats it as two separate themes and creates two theme tables. The theme table you've just opened wasn't affected by the join operation in the preceding exercise.

Notice that the theme table contains no information about recreation areas.

3.   Make the Project window active by selecting *ex16b.apr* from the Window menu. Click on the Tables icon if it isn't already highlighted. Click the Add button. In the Add Table dialog box, select the drive where you loaded the data for this book, then find *\gtkav\data\ch16* in the Directories list. Click on "rec.dbf" to select it, then click OK to add it to the project.

The table contains a record for every recreation area in California, which means that for each county in the theme table, there may be zero, one, or *many* matching records. If you were to join the tables, ArcView would find the first matching record for each county, append its attributes to the theme table, and ignore any further matching records. To preserve the one-to-many relationship between the two tables, you'll link them instead.

4. Move the rec.dbf table to the upper left corner of the ArcView window. Click on the Cnty_name field to make it active.

5. Make the theme table active. (Select it from the Window menu if you can't see it.) Move it to the lower right corner of the ArcView window. Click on the Name field to make it active. You can resize the tables if you like.

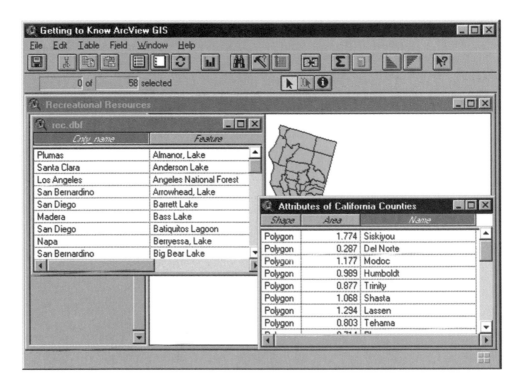

6. From the Table menu, select Link. A one-way link is established from the theme table (destination) to the rec.dbf table (source). Both tables remain open.

7.   In the theme table, click on the record for Modoc County (it's the third from the top). All Modoc County records are selected in the rec.dbf table and Modoc County is highlighted in the view.

   8.   Make the rec.dbf table active, then click the Promote button to move the selected records to the top of the table.

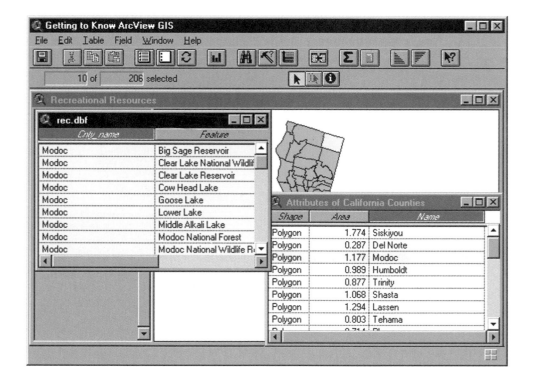

You see a list of recreation areas located in Modoc County. Selecting one record in the theme table caused all corresponding records to be selected in the rec.dbf table.

Linking tables in one and two directions. Links are unidirectional: if you select a record in the destination table, the corresponding records in the source table are also selected, but if you select a record in the source table, no records are selected in the destination table. You can establish a two-way link by reversing the source and destination tables and linking them again. When a two-way link is established, clicking on a record in either table will highlight corresponding records in the other table. Similarly, querying either table will select features in the view. For more information, search for these Help Topics: *Link, Linking tables.*

You can't classify the theme on fields in the linked rec.dbf table. To classify the counties by number of recreation areas, you would need to perform some additional steps. First, you would summarize the rec.dbf table on the Cnty_name field. This would create a summary table with a "Count" field containing the number of recreation areas in each county. (For a review of the Summarize function, see chapter 15.)

Next, you would join the summary table (source table) to the theme table (destination table), using the Name field as the common field. After joining the tables, you would use the Legend Editor to classify the counties on the values in the Count field (as you did in exercise 16a). The resulting view would shade the counties according to the number of recreation areas in each.

9.   Close the two open tables.

You're well on your way to your goal of finding the best California counties to live in. You've evaluated two criteria: income and recreation areas. You can consider additional criteria (for instance, educational opportunities, employment, and housing prices) by joining or linking different tables to the theme table.

Creating hot links

With ArcView's *hot link* feature, you can link a feature in a view to a data source such as a text file, image, ArcView document, or Avenue script. When you click on the feature with the Hot Link tool, ArcView opens the linked data source. You can even hot link to an Avenue script that starts another program or plays video clips (you'll learn more about Avenue, ArcView's programming language, in chapter 27).

To create a hot link, you add a new field to the theme table and enter the names of the data sources to be hot linked. Each data source name thus becomes an attribute of a particular record; the data source is hot linked to the feature associated with that record. Next, you use the hot link theme properties to tell ArcView the name of the new field and what kind of data source is being hot linked (for example, image files, ArcView documents, or text files).

Suppose you're a biologist working with a conservation agency in Brazil. You've just returned from a trip to the coastal rain forests in the northeastern part of the country, and you're going to give a presentation on ecological efforts under way in the region. To make your presentation more interesting, you'll show a map of the area, along with photographs of animals you encountered on your trip. You'll hot link points on the map to your photos. When you click on a point, ArcView will display a photo.

Exercise 16c

1. From the File menu, choose Open Exercise. In the Exercises scrolling list, select "ex16c," then click OK.

The view, Photo sites in the Brazilian rain forest, shows a map of the coastal rain forest area you visited. Two point themes, Mammal sites and Bird sites, show the locations where you took photos.

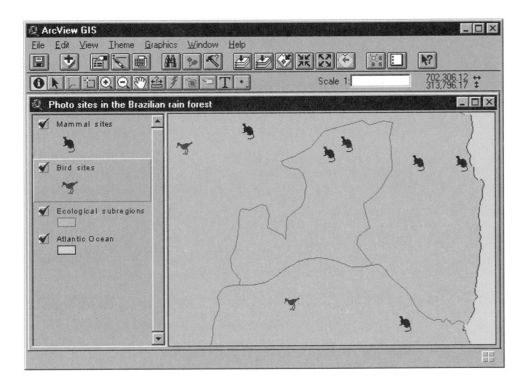

2. From the Window menu, choose *ex16c.apr* to make the Project window active. You see that the project contains several views.

In addition to the Photo sites view, there are various views containing pictures of animals. In each of these views, a scanned photograph has been loaded as an image theme (see chapter 8) and the Table of Contents has been hidden. These views will be hot linked to the features in the point themes.

3.   Double-click on "Photo sites in the Brazilian rain forest" to make the view active.

You'll hot link the two point features in the Bird sites theme to the two photographs of birds. To do this, you'll add a new field to the Bird sites theme table and enter the names of the views that contain the bird photos. Then you'll set the hot link theme properties.

4.   With the Bird sites theme active, click the Open Theme Table button. The Attributes of Bird sites table appears.

You'll add a field called "Photo" to this table.

5.   From the Table menu, choose Start Editing. The field names become non-italic, indicating that the table is now editable.

6. From the Edit menu, choose Add Field. The Field Definition dialog box appears. In the Name field, type **Photo.** From the Type drop-down list, select "String." In the Width field, type **40.**

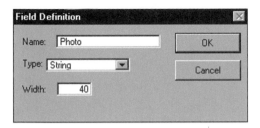

7. Click OK to create the new field in the theme table.

Because you're hot linking features to ArcView documents within the project, you need only enter the document name. (To hot link to files not contained within your project, you must specify the path name as well as the file name.)

8. Click on the Edit tool. In the Photo field, click in the first record and type **Brazilian tanager.** (Spelling matters, but case doesn't.) Press Enter to commit your edit.

9. Type **Northeastern macuco** in the next record. Press Enter. The attribute table looks like this:

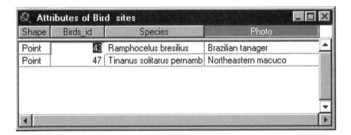

10. Choose Stop Editing from the Table menu. The Stop Editing dialog box appears. Choose Yes to save your edits.

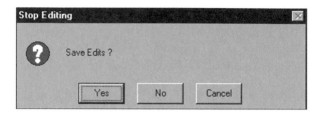

11. Close the theme table.

Now you'll tell ArcView that the Photo field contains the hot link information and that items in the Photo field are ArcView documents.

12. From the Theme menu, select Properties to open the Theme Properties dialog box. Scroll down and click on the Hot Link icon (along the left margin) to display the hot link property options.

13. From the Field drop-down list, choose "Photo." This specifies the field in the theme table that contains the hot link information. From the Predefined Action drop-down list, choose "Link to Document" (in this case, the document is a view).

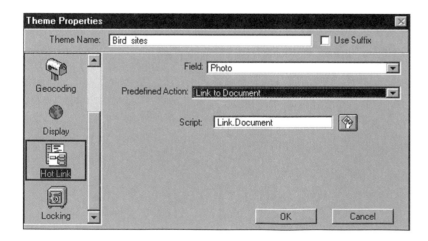

A default script (program) called "Link.Document" is automatically selected. This script contains the instructions ArcView needs to open a document when a hot-linked feature is clicked.

14. Click OK to set the hot link properties.

Now you'll test the hot links.

15. Make sure the Bird sites theme is active and click on the Hot Link tool. (This tool is enabled only when hot link properties have been set for the active theme.) The cursor changes to a lightning bolt as you move it over the view. Place the tip of the lightning bolt on the bird site in the upper left corner of the view and click. The view containing a photo of the Northeastern macuco opens. (If the view doesn't open on the first click, move the tip of the lightning bolt closer to the center of the marker and try again.)

16. Close the view of the photograph and click on the other bird site. A view with a photo of the Brazilian tanager opens.

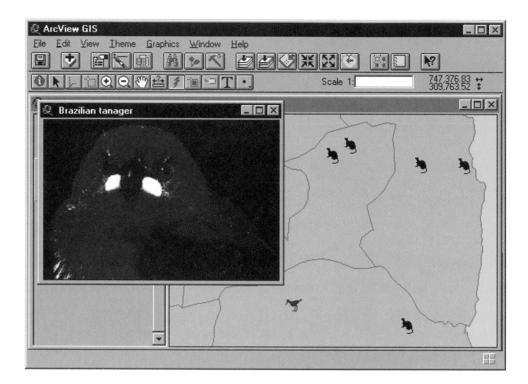

You've completed the hot links for the Bird sites theme. (The hot links for the mammal photo sites have already been done.)

17. Close the view of the Brazilian tanager. Make the Mammal sites theme active by clicking on it in the Table of Contents. With the Hot Link tool, click on a mammal photo site. Close the photo view and click on a few of the other mammal photo sites.

Now you're ready to make your presentation, complete with maps and photographs.

If you want to go on to the next chapter, leave ArcView running. Otherwise, choose Exit from the File menu.

SECTION 5

Analyzing spatial relationships

In the next four chapters you'll learn how to analyze features based on where they're located in relation to other features. In chapters 17, 18, and 19, you'll use theme-on-theme selection to find features that are nearby, adjacent to, inside, or intersecting other features in the same theme or in different themes. In chapter 20, you'll use spatial join to append the attributes of features in one theme to features in another theme based on their locations.

SECTION 5:
Analyzing spatial relationships

Finding the features nearby

Finding points near lines
Finding points near other points
Finding adjacent features

Finding the features nearby

When you analyze the relationships between map features, you might need to know which features are within a certain distance of other features or are adjacent to other features. ArcView GIS uses *theme-on-theme selection* to analyze the locations of features in relation to other features, whether in the same theme or in different themes. In this chapter, you'll focus on two types of feature relationships, *proximity* (the distance between features) and *adjacency* (features that share the same boundary).

Finding points near lines

Suppose you're interested in buying a gas station near Interstate 40 in Old Town, New Mexico. One of your requirements is that it must be within 1,000 feet of the interstate so you can attract as many freeway drivers as possible. Using ArcView's theme-on-theme selection, you can find out which gas stations meet your criteria.

Exercise 17a

1. If necessary, start ArcView. From the File menu, choose Open Exercise. In the Exercises scrolling list, select "ex17a," then click OK. When the project opens, you see a view with two themes: a line theme, Streets, and a point theme, Stations.

ENVIRONMENTAL SYSTEMS RESEARCH INSTITUTE, INC.

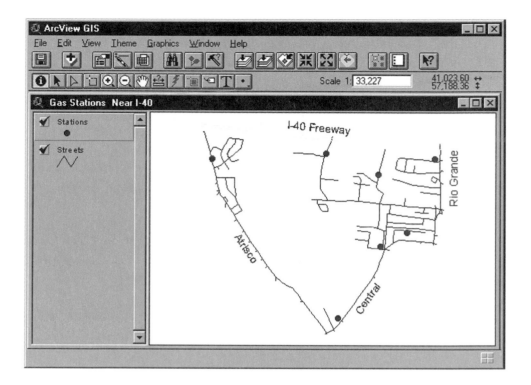

The Stations theme is active and the I-40 freeway is selected (shown in yellow) in the Streets theme. You'll use this feature to select gas stations within 1,000 feet of it.

The Stations theme is the *target* theme. Features in this theme will be selected by features in the Streets theme. (The target theme must be active to perform a theme-on-theme selection.)

The Streets theme is the *selector* theme. That is, features in this theme will be used to select features in the Stations theme.

If features are highlighted in the selector theme, ArcView uses only the highlighted features to select features in the target theme.

---

2. From the Theme menu, choose Select By Theme. The Select By Theme dialog box displays.

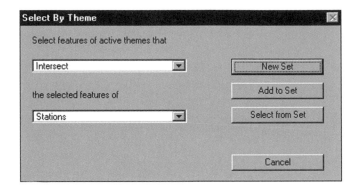

In this dialog box, you define a spatial analysis operation with choices from the drop-down lists. For example, the default choices form this sentence: Select features of active themes that intersect the selected features of Stations. (This isn't what you want, so you'll make different choices from the drop-down lists.)

Because the selector theme's feature type determines what types of spatial relationships you can analyze, you'll choose the selector theme first.

3. In the Select By Theme dialog box, select "Streets" from the lower drop-down list.

4. Choose "Are Within Distance Of" from the upper drop-down list.

5. Type **1000** feet as the Selection distance. (The selection distance units match the Distance Units setting in the View Properties. See chapter 11 for a review of setting distance units.)

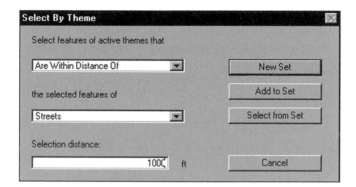

Your choices form this sentence: Select features of active themes that are within distance of the selected features of Streets. The specified selection distance is 1,000 feet. Therefore, points in the Stations theme (the target theme) will be highlighted if they are within 1,000 feet of the I-40 freeway (the highlighted feature in the selector theme).

6.   Click New Set. ArcView finds two gas stations within 1,000 feet of the I-40 freeway.

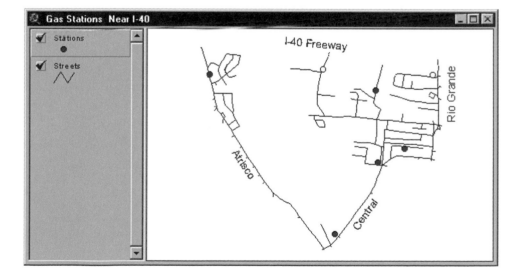

To find out more about the selected stations, you'll use the Identify tool.

7.   Click on the Identify tool if it's not already highlighted. Then click on each selected station to display its attributes in the Identify Results dialog box.

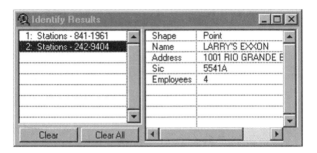

The selected stations are Larry's Exxon and Ann's Mart Station #1963. You'll analyze these locations in the next exercise.

8.   Close the Identify Results dialog box.

If you want to go on to the next exercise, leave the project open.

Finding points near other points

Now you know that two gas stations, Larry's Exxon and Ann's Mart Station #1963, are within 1,000 feet of I-40. Unfortunately, neither is currently for sale. You'd like to make a tempting offer, but first you must know the market potential of each station. Because you'd like to sell gasoline to business customers and employees as well as freeway travelers on I-40, you'll use theme-on-theme selection to find out which gas station is closer to more businesses.

Exercise 17b

1.   If *ex17a.apr* is open, close the Gas Stations Near I-40 view and open the Businesses Near Ann's Mart Station #1963 view from the Project window. Otherwise, choose Open Exercise from the File menu. In the Exercises scrolling list, select "ex17b," then click OK. When the

project opens, you see a view with two themes: Business and Streets. The Streets theme is active, and Ann's Mart Station #1963 is selected in the Business theme.

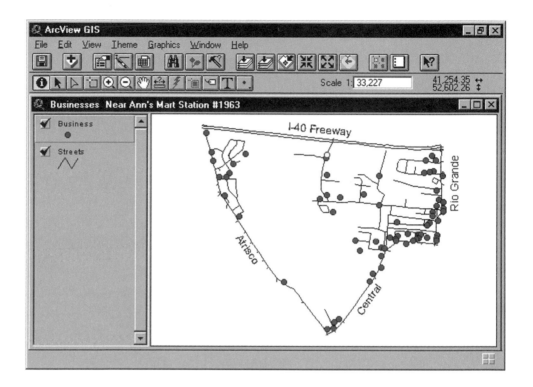

You'll use theme-on-theme selection to select businesses in the Business theme that are within a quarter mile (1,320 feet) of Ann's Mart Station #1963. In this case, you'll use the Business theme as both target and selector theme.

2.   Make sure the Business theme is active to make it the target theme. Choose Select By Theme from the Theme menu. The Select By Theme dialog box displays.

3.   Select "Business" from the lower drop-down list to make it the selector theme.

4.   From the upper drop-down list, select "Are Within Distance Of."

5. Type **1320** feet as the Selection distance.

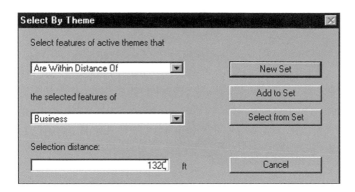

Your choices form this sentence: Select features of active themes that are within distance of the selected features of Business. The specified selection distance is 1,320 feet. Because the Business theme is both the target and selector theme, points in the theme that are within 1,320 feet of the highlighted point (Ann's Mart Station #1963) will also be highlighted.

6. Click New Set. ArcView selects businesses within 1,320 feet of Ann's Mart Station, which is also selected.

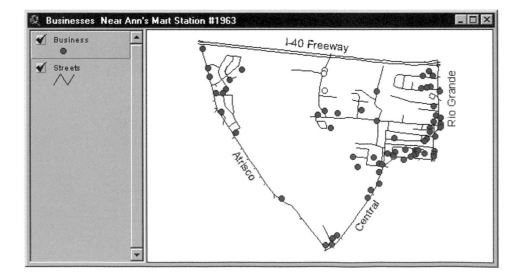

To see which businesses ArcView selected and examine their attributes, you'll open the theme table.

7. Click the Open Theme Table button to open the attribute table for the Business theme. The status portion of the Table tool bar shows that nine businesses are selected. Click the Promote button to move the selected records to the top of the table.

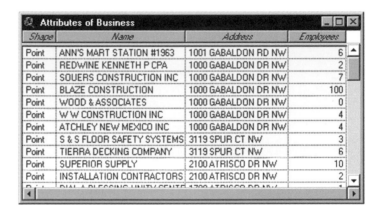

Shape	Name	Address	Employees
Point	ANN'S MART STATION #1963	1001 GABALDON RD NW	6
Point	REDWINE KENNETH P CPA	1000 GABALDON DR NW	2
Point	SOUERS CONSTRUCTION INC	1000 GABALDON DR NW	7
Point	BLAZE CONSTRUCTION	1000 GABALDON DR NW	100
Point	WOOD & ASSOCIATES	1000 GABALDON DR NW	0
Point	W W CONSTRUCTION INC	1000 GABALDON DR NW	4
Point	ATCHLEY NEW MEXICO INC	1000 GABALDON DR NW	4
Point	S & S FLOOR SAFETY SYSTEMS	3119 SPUR CT NW	3
Point	TIERRA DECKING COMPANY	3119 SPUR CT NW	6
Point	SUPERIOR SUPPLY	2100 ATRISCO DR NW	10
Point	INSTALLATION CONTRACTORS	2100 ATRISCO DR NW	2

Excluding Ann's Mart Station, there are eight selected businesses with 126 employees. Many of the businesses have the same address, indicating that they are located in a shopping center.

8. Close the Attributes of Business table.

Next you'll examine the relationship of Larry's Exxon to surrounding businesses.

9. Close the current view, then open the Businesses Near Larry's Exxon view from the Project window. You see two themes: Business and Streets. Streets is active and Larry's Exxon is selected in the Business theme.

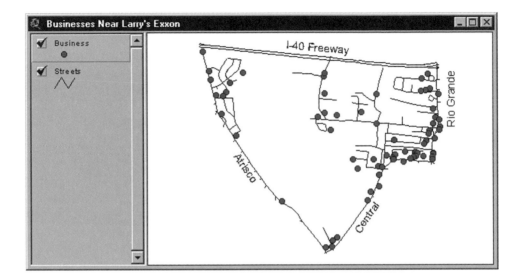

10. Make sure the Business theme is active (it will be the target theme). Choose Select By Theme from the Theme menu. The Select By Theme dialog box displays.

11. Select "Business" from the lower drop-down list. The Business theme will be the selector as well as the target theme.

12. From the upper drop-down list, select "Are Within Distance Of."

13. Type **1320** feet as the Selection distance.

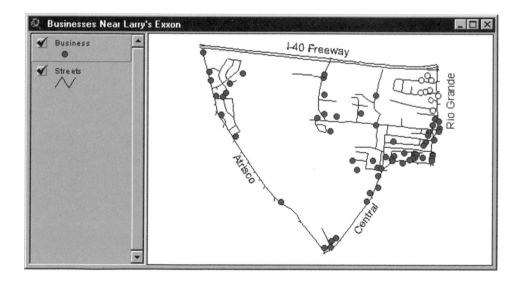

Your choices form this sentence: Select features of active themes that are within distance of the selected features of Business. The specified selection distance is 1,320 feet. Because the Business theme is both the target and selector theme, the points in the theme that are within 1,320 feet of the highlighted point will also be highlighted.

14. Click New Set. ArcView selects businesses within a quarter mile of Larry's Exxon.

Now you can examine the attributes of these selected businesses.

15. Click the Open Theme Table button to open the attribute table for the Business theme. Click the Promote button to move the selected records to the top.

Shape	Name	Address	Employees
Point	RIO GRANDE INN	1015 RIO GRANDE BLVD	120
Point	LARRY'S EXXON	1001 RIO GRANDE BLVD	4
Point	NORM'S WRECKING COMPANY	2328 ASPEN AVE NW	2
Point	ALBUQUERQUE SOUTHWEST A	903 RIO GRANDE BLVD	7
Point	MAXIE ANDERSON-BEN ABRUZ	2435 ZEARING AVE NW	3
Point	ANDERSON SOUTHWEST	2437 ZEARING AVE NW	8
Point	MILLER ERIC	2437 ZEARING AVE NW	4
Point	ADVANTAGE APPLIANCES	830 RIO GRANDE BLVD	11
Point	AMIGO RENTALS	830 RIO GRANDE BLVD	12
Point	BLAKES LOTA BURGER #70	777 RIO GRANDE BLVD	16
Point	SUPERIOR SUPPLY	2100 ATRISCO DR NW	10

Excluding Larry's Exxon, there are nine selected businesses, with 183 employees.

Not only is Larry's Exxon closer to more businesses with more employees than Ann's Mart Station, but it's conveniently near the Rio Grande Inn, a good business to have near a gas station. Now that you know something about the market potential of Larry's Exxon, you're ready to make an offer.

Finding adjacent features

ArcView can also find features adjacent to selected features. To see how this works, consider a situation involving parcels of land. Suppose that Old Town is planning major improvements to its drainage system. Municipal planners need answers to these questions: Which parcels are part of or adjacent to drainage ditches? How many of these parcels are in the city's jurisdiction and how many in the county's? What's the total acreage of parcels by jurisdiction?

Exercise 17c

1. From the File menu, choose Open Exercise. In the Exercises scrolling list, select "ex17c," then click OK. When the project opens, you see a view with one theme, Zoning. The Zoning theme is active and the parcels zoned for drainage ditches are selected.

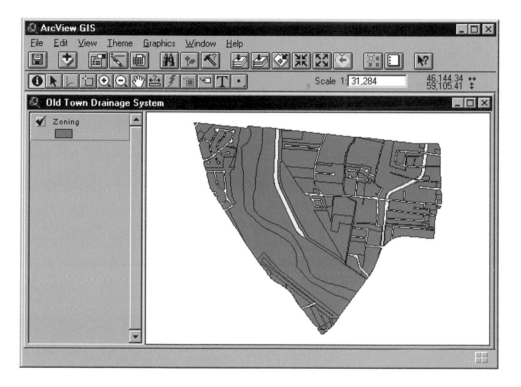

First you'll find out which parcels are adjacent to drainage ditches.

2. From the Theme menu, choose Select By Theme to display the Select By Theme dialog box.

3. Select "Are Within Distance Of" from the upper drop-down list. ("Zoning" is already selected in the lower drop-down list, as it's the only theme in the view.)

The Selection distance box is set to 0 feet by default. Because you want to find only the parcels adjacent to drainage ditches, you'll leave the selection distance as it is.

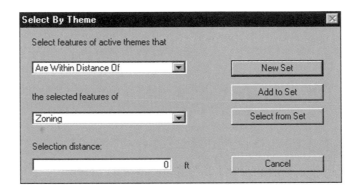

Your choices form this sentence: Select features of active themes that are within distance of the selected features of Zoning. The selection distance is 0 feet. The Zoning theme is both the target and the selector theme. Features that touch highlighted features in the theme will also be highlighted.

4. Click New Set. ArcView selects parcels adjacent to the highlighted parcels of the drainage system.

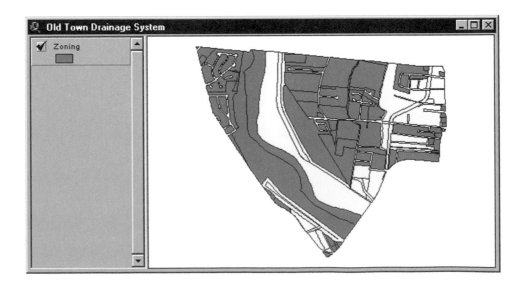

To find out how many parcels are selected, you'll open the theme table.

5. Click the Open Theme Table button, then click the Promote button to move the selected records to the top.

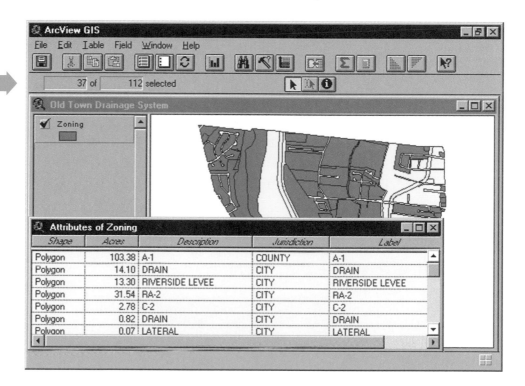

In the tool bar, you can see the number of selected records in the table. These are the records for the parcels that are adjacent to or part of the drainage system. The new drainage improvement program will affect 37 parcels.

Next you'll determine the total acreage of these parcels by jurisdiction, using Summarize.

6. In the Attributes of Zoning table, click on the Jurisdiction field title to make the field active.

 7. Click the Summarize button to display the Summary Table Definition dialog box.

8. From the upper drop-down list (Field), select "Acres." From the lower drop-down list (Summarize by), select "Sum."

9. Click Add to add your selections to the Summary statistics box.

10. Click the Save As button to navigate to the *drive:\directory* where you want to save the new table and name it **acre_sum.dbf.**

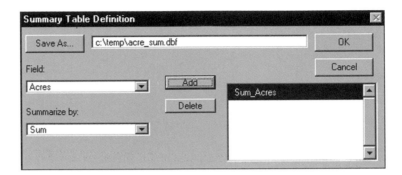

11. Click OK. ArcView sums the acres of selected parcels by jurisdiction and places the results in a new table.

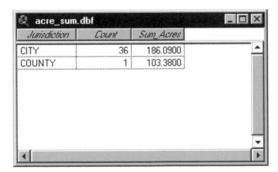

The total number of city parcels adjacent to the drainage system is 36, with a total acreage of 186.09. Only one county parcel, with an acreage of 103.38, is adjacent to the drainage system.

Now Old Town knows how much city and county land the new drainage improvement program will affect.

If you want to go on to the next chapter, leave ArcView running. Otherwise, choose Exit from the File menu.

SECTION 5:
Analyzing spatial relationships

Finding the features within

Finding points within polygons

Finding polygons within polygons

Finding the features within

Using theme-on-theme selection in ArcView GIS, you can find the points, lines, and polygons in one theme that fall completely within a polygon or polygons in another theme. Conversely, you can find polygons in one theme that contain particular points, lines, or polygons in another theme. This type of spatial relationship—features inside other features—is known as *containment*. Finding out whether a feature is inside or outside a boundary can be crucial to making decisions.

Finding points within polygons

Your company is transferring you to the Atlanta region, and you'd like to purchase a home after you get there. Before you call a real estate agent long distance (you're on the West Coast now), you'd like to become familiar with the region yourself, and possibly even identify some areas you might like to live in. You're primarily interested in areas where population is low compared to the rest of the region, and more importantly, areas where your thirteen-year-old daughter can attend middle school close to home.

Your company uses ArcView and you have access to a copy, along with the sample data that comes with it. The data includes demographic information you can use to study the population characteristics of the region, but no information about schools. Conveniently, one of your old college chums works for a regional planning agency in Georgia, and he's willing to send you some data containing the locations and names of public school facilities in the Atlanta region. You'll use these data sets and ArcView to find suitable areas to live.

Exercise 18a

1. If necessary, start ArcView. From the File menu, choose Open Exercise. In the Exercises scrolling list, select "ex18a," then click OK. When the project opens, you see a view with three themes: Middle Schools, Census Tracts, and Counties.

The Middle Schools theme contains the locations of all the middle schools in the region. The Census Tracts theme contains the census tracts in the region along with demographic information (i.e., per capita income, total population, population growth, age characteristics, and more). The Counties theme contains the boundaries of four counties in the region. You don't see it because ArcView draws the other two themes on top of it.

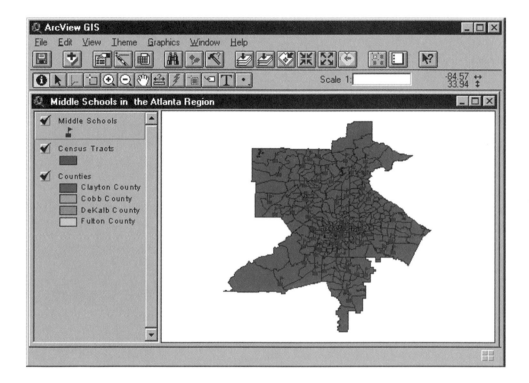

You'll use the Census Tracts theme to find areas with low population.

 2.   Make the Census Tracts theme active, then click the Open Theme Table button to open the Attributes of Census Tracts table.

When the table opens, you see fields containing demographic information for each census tract. The Pop_90 field contains the 1990 population value for each tract.

3. Click on the Pop_90 field to make it active.

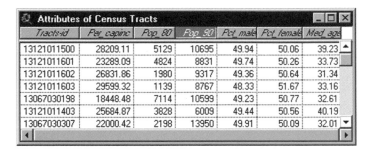

4. From the Field menu, choose Statistics. ArcView displays statistics for the Pop_90 field.

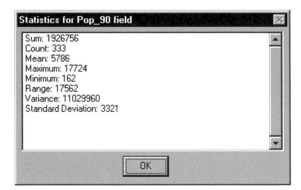

In the dialog box that displays, you see that the mean population for all the census tracts is 5,786. Because you want to live in an area with low population compared to the rest of the region, you'll start by finding census tracts that have a population that's less than the mean. You'll use the Query Builder to find these areas.

5. Click OK to close the Statistics dialog box, then close the Attributes of Census Tracts table.

6. Click the Query Builder button. The Query Builder dialog box displays.

7. In the Fields list, double-click on "[Pop_90]," click the "<" button, then type **5786** in the query text box. (By default, ArcView encloses the query in parentheses.)

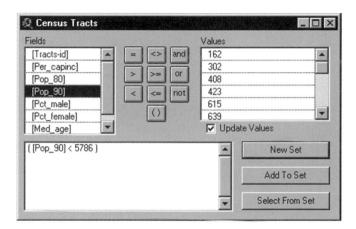

8. Click the New Set button. ArcView selects and highlights the census tracts that have a population less than 5,786. Close the Query Builder.

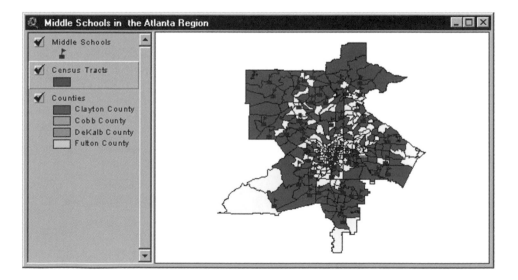

By looking at the view, you can see that some of the highlighted census tracts have middle schools. You want to find census tracts that have low population and also contain middle schools. Since the census tracts and middle schools are in separate themes, you'll use theme-on-theme selection to find census tracts that meet both of your criteria.

The Middle Schools theme is the selector theme. That is, features in this theme will be used to select features in the Census Tracts theme.

9. With the Census Tracts theme active, choose Select By Theme from the Theme menu. The Select By Theme dialog box displays.

The Census Tracts theme is the target (active) theme. Features in this theme will be selected by features in the Middle Schools theme.

10. From the lower drop-down list, choose "Middle Schools." From the upper drop-down list, choose "Completely Contain." Your selections form this sentence: Select features of active themes that completely contain the selected features of Middle Schools.

Because no features are selected in the Middle Schools theme, ArcView uses all the theme's features to find and select features in the Census Tracts theme.

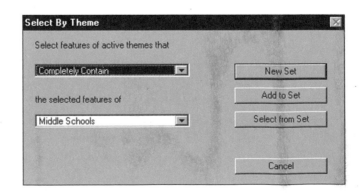

ENVIRONMENTAL SYSTEMS RESEARCH INSTITUTE, INC.

Because you've already selected a set of census tracts that meet your first criterion (low population), you want ArcView to select census tracts from this set.

11. Click the Select from Set button. ArcView selects census tracts from the currently selected set that also contain middle schools.

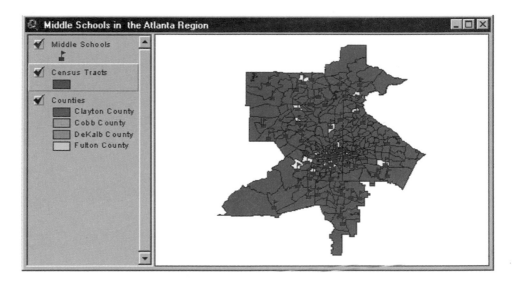

ArcView highlights those census tracts with a population less than 5,786 and at least one middle school. These areas meet both of your criteria for areas you'd like to live in.

If you want to go on to the next exercise, leave the project open.

Finding polygons within polygons

Suppose you've just learned from a future coworker in the Atlanta region that property taxes are lower in Cobb County than in other counties in the region. You want to use ArcView to find out which of the census tracts you selected are in Cobb County. This may help you narrow your search for a place to live. Once you narrow your choice of areas, you'll want to get information about the middle schools located there.

Exercise 18b

1. If *ex18a.apr* is open, continue. Otherwise, choose Open Exercise from the File menu. In the Exercises scrolling list, select "ex18b," then click OK. When the project opens, you see three themes: Middle Schools, Census Tracts, and Counties. (You don't see the Counties theme because ArcView draws the other two themes on top of it.) The census tracts you selected in the last exercise are highlighted in the Census Tracts theme.

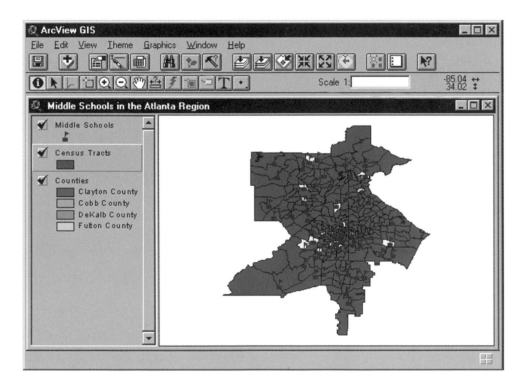

To find out which of the highlighted census tracts are in Cobb County, you first need to select Cobb County in the Counties theme. To make the Counties theme visible, you'll turn off the other themes.

2. Click on the check boxes for the Middle Schools and Census Tracts themes to turn them off. The Counties theme is now visible.

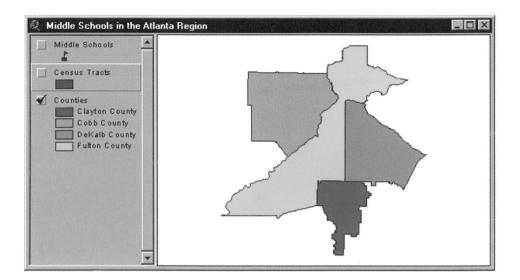

You can see from the legend that the counties are classified according to their names. You'll use the Select Feature tool to select the feature (polygon) that represents Cobb County.

 3. Make the Counties theme active, click on the Select Feature tool, then click on Cobb County. ArcView selects and highlights this county.

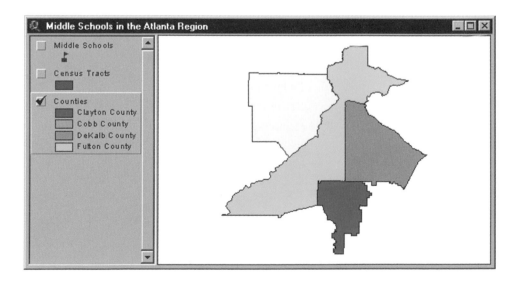

Now you want to see the Census Tracts and Middle Schools themes, so you'll turn them back on.

4. Click on the check boxes for the Census Tracts and Middle Schools themes to turn them on. These themes draw on top of the Counties theme.

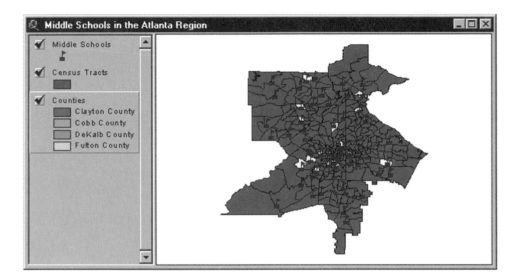

To find out which selected census tracts are inside Cobb County, you'll use theme-on-theme selection.

5. Make the Census Tracts theme active to make it the target theme.

6. From the Theme menu, choose Select By Theme to display the Select By Theme dialog box. From the lower drop-down list, select "Counties" to make it the selector theme. From the upper drop-down list, select "Are Completely Within." Your selections form this sentence: Select features of active themes that are completely within the selected features of Counties.

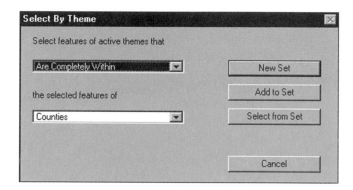

Because you've already selected a set of census tracts that meet two of your criteria (low population and at least one middle school), you want ArcView to select census tracts from this set.

7.   Click the Select from Set button. ArcView selects census tracts from the currently selected set that are completely inside Cobb County.

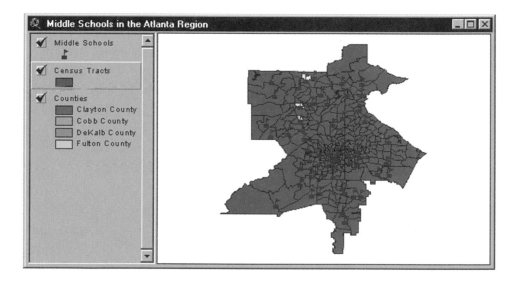

These census tracts meet all your criteria. They have a population that's less than the average for the region, they have middle schools, and they're in Cobb County, where property taxes are lower.

Now that you've narrowed down the number of possible areas to move to, your final task is to get information about their middle schools. You'll select these schools using theme-on-theme selection, then examine their attributes to get their names.

8. Make the Middle Schools theme active to make it the target theme.

9. From the Theme menu, choose Select By Theme to display the Select By Theme dialog box. From the lower drop-down list, select "Census Tracts" to make it the selector theme. From the upper drop-down list, choose "Are Completely Within." Your selections form this sentence: Select features of active themes that are completely within the selected features of Census Tracts.

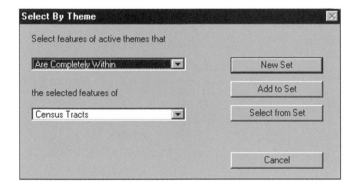

10. Click the New Set button. ArcView selects the middle schools that are inside the selected census tracts. (Selected census tracts and selected middle schools are highlighted in the view.)

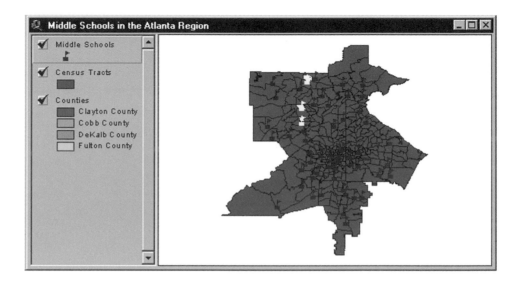

    11. Click the Open Theme Table button to open the Attributes of Middle Schools table, then click the Promote button to move the selected schools to the top of the table.

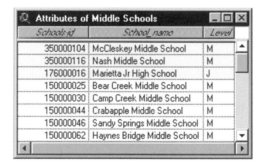

You now have the names of the schools in the areas that meet your criteria. You can contact the schools to ask when the next session begins, whether they operate year-round, what the average class size is, and so on. Answers to these questions may help you choose a place to live. Then you can call an agent and start looking for a home.

If you want to go on to the next chapter, leave ArcView running. Otherwise, choose Exit from the File menu.

Finding features that intersect other features

Finding lines that intersect other lines

Finding polygons that intersect other polygons

Finding features that intersect other features

When features share the same geographic space, they overlap, or *intersect*. ArcView GIS can find features that intersect, whether they're in the same theme or in different themes. Using theme-on-theme selection, you can find and select lines that intersect other lines or polygons, and polygons that intersect other polygons.

Once you find and select intersecting features, you can perform other ArcView operations on them and on their attributes.

Finding lines that intersect other lines

Suppose that as part of an emergency planning strategy for a large city, the city's emergency preparedness committee is studying major earthquake faults in its suburban areas. One objective of the study is to develop a plan to deal with potential flooding and health risks associated with the rupture of water and sewer lines in the event of a major quake. As the first step in the study, the committee uses ArcView to find and select those water and sewer pipelines that intersect faults.

Exercise 19a

1. If necessary, start ArcView. From the File menu, choose Open Exercise. In the Exercises scrolling list, select "ex19a," then click OK. When the project opens, you see a view with two line themes, Pipelines and Faults, and one polygon theme, Basin.

ENVIRONMENTAL SYSTEMS RESEARCH INSTITUTE, INC.

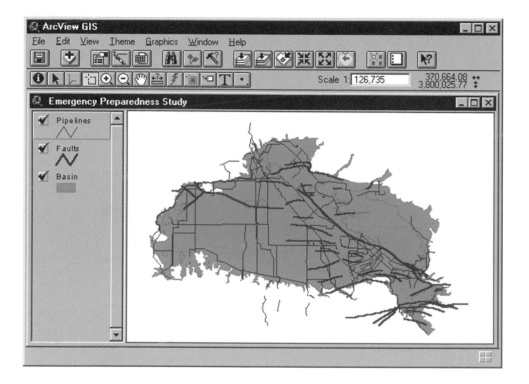

To find out which pipelines intersect faults, you'll use theme-on-theme selection.

The Pipelines theme is the target (active) theme. Features in this theme will be selected by features in the Faults theme.

2.   Make sure the Pipelines theme is active. Choose Select By Theme from the Theme menu. The Select By Theme dialog box displays.

3. Select "Faults" from the lower drop-down list. Then choose "Intersect" from the upper drop-down list.

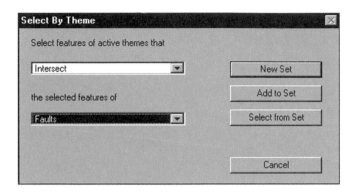

Your choices form this sentence: Select features of active themes that intersect the selected features of Faults. (Because none of the features in the Faults theme is selected, all features in the theme will be used for the analysis.)

4. Click New Set. ArcView selects pipelines that intersect faults.

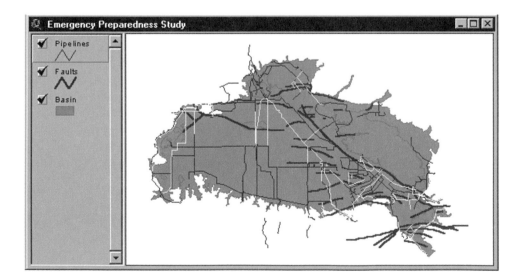

Now that the vulnerable pipelines are selected, the committee wants to determine their total number and length so they can estimate potential repair and replacement costs. Because the pipelines belong to more than one agency, they'll break down the totals by agency.

5. With the Pipelines theme active, click the Open Theme Table button. The Attributes of Pipelines table opens.

To see the selected records better, you promote them.

6. Click the Promote button. Records for the selected pipelines now appear at the top of the table.

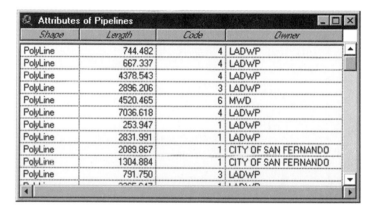

Now you'll use the Summarize function to calculate the total number and length of these pipelines according to owner.

7. In the Attributes of Pipelines table, click on the Owner field to make it active. ArcView will use the unique values in this field to summarize the selected records in a new table.

8. Click the Summarize button. The Summary Table Definition dialog box displays.

9. From the upper drop-down list (Field), select "Length." From the lower drop-down list (Summarize by), select "Sum." Then click the Add button. This adds your selection to the Summary statistics box on the right. Click the Save As button to navigate to the *drive:\directory* where you want to save the new table ArcView creates and call it **owlength.dbf.**

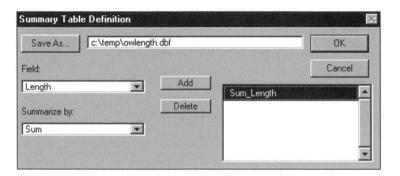

10. Click OK to create the summary table.

ArcView creates one record for each unique owner it finds. The new table shows there are five owners of the selected pipelines. The Count field reports the total number of pipelines for each owner, and the Sum_Length field lists their total length.

Now the committee can use this information as the starting point for working with the responsible agencies to develop a plan to deal with potential flooding and health risks.

Finding polygons that intersect lines. You can also use theme-on-theme selection to find and select polygons that intersect lines in another theme. For example, suppose you have one theme containing polygons that represent businesses and another containing lines that represent earthquake faults. You want to find out which businesses are vulnerable to earthquakes, so you perform a theme-on-theme selection. The theme containing businesses is the target theme; the theme containing earthquake faults is the selector theme. ArcView uses the earthquake fault lines to select businesses (polygons) they intersect. For more information, search for these Help Topics: *Select By Theme, Select By Theme (Dialog box), Spatial relation types, Theme on theme selection.*

Finding polygons that intersect other polygons

Like earthquakes, floods require emergency planning and special programs for protection. Legislation requires communities in flood-prone areas to participate in flood insurance programs. These programs, aimed at protecting property owners, require owners to purchase flood insurance at government-subsidized rates.

As a participant in a flood insurance program, one city located on a major floodplain developed a database containing the boundaries of 100-year and 500-year flood zones. (These zones are expected to flood at least once during the designated time span.) The city wants to notify all property owners within the 100-year zone about special low-cost loans for elevating structures above the base flood level, thereby cutting insurance costs.

Using ArcView's theme-on-theme selection, the city can determine which land parcels are located within the 100-year flood zone and then notify the owners about the loan program.

Exercise 19b

1. From the File menu, choose Open Exercise. In the Exercises scrolling list, select "ex19b," then click OK. When the project opens, you see a view with two themes: Floodzones and Parcels. The 100-year flood zone is selected in the Floodzones theme.

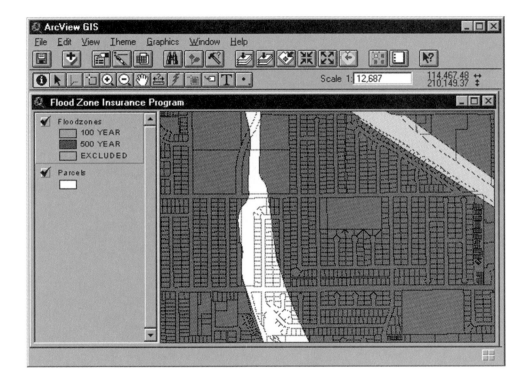

To see all the features in both themes, you'll zoom out.

2. Click the Zoom to Full Extent button.

Next you'll find out which parcels intersect the 100-year flood zone.

3. Make the Parcels theme active to make it the target theme and choose Select By Theme from the Theme menu. The Select By Theme dialog box displays.

4. Choose "Floodzones" from the lower drop-down list to make it the selector theme if it is not already selected. Then, from the upper drop-down list, choose "Intersect."

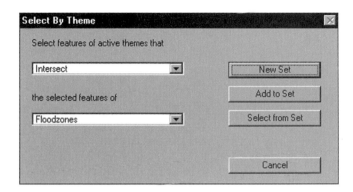

Your choices form this sentence: Select features of active themes that intersect the selected features of Floodzones. The Parcels theme is the active theme and the 100-year flood zone is the selected feature in the Floodzones theme.

5. Click New Set. ArcView selects parcels that intersect the 100-year flood zone.

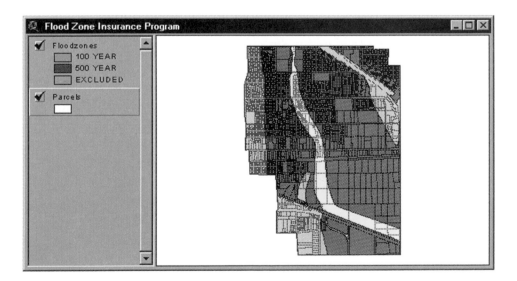

To see the selected parcels better, you'll turn off the Floodzones theme.

6. In the Table of Contents, click on the check box next to the Floodzones theme to turn it off. ArcView redraws the Parcels theme.

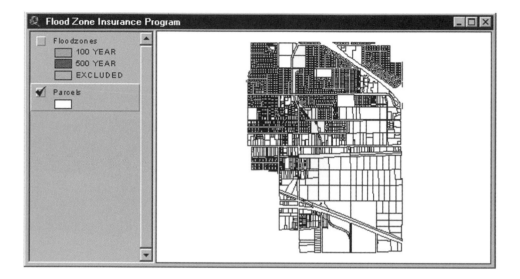

Next you'll open the theme table and examine the selected records.

7. With the Parcels theme active, click the Open Theme Table button, then click the Promote button. The selected records display at the top of the table. Scroll to the right to examine the attributes.

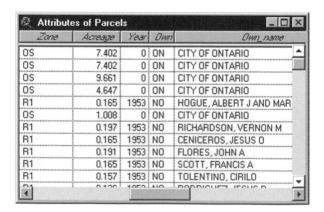

The parcel attributes include assessor's parcel number, zoning codes, General Plan codes, acreage, owner name, owner address, and more. The city now has a list of all parcels that would be affected by a 100-year flood as well as the information it needs to contact the owners about the special loan program.

If you want to go on to the next chapter, leave ArcView running. Otherwise, choose Exit from the File menu.

Finding features that don't intersect. Sometimes you need to find features that *don't* intersect other features. For example, suppose you want to find the parcels that are outside of the 100-year flood zone. You use ArcView's theme-on-theme selection to find and select the parcels that intersect the flood zone, then use ArcView's Switch Selection function to toggle the selected set from the parcels that do intersect (unselects these) to the parcels that *don't* intersect (selects these instead). For more information, search for this Help Topic: *Switch Selection.*

SECTION 5:
Analyzing spatial relationships

Spatially joining tables

Joining attributes based on containment

Joining attributes based on proximity

CHAPTER 20

Spatially joining tables

When you use theme-on-theme selection, ArcView GIS uses the features in one theme to find and select features (and their attributes) in another theme. The selected features and attributes remain in separate themes.

In another type of spatial analysis, called *spatial join,* ArcView appends the fields of one theme table to those of another theme table, based on the locations of features in the two themes with respect to each other.

ArcView uses two kinds of spatial relationships to analyze the locations of features in two separate themes: *nearest* and *inside.* Whenever ArcView finds features that satisfy one of these spatial relationships, the attributes of features in one theme are appended to those in the other theme.

The relationship (*nearest* or *inside*) ArcView uses to compare feature locations depends on the types of features in the two themes. For example, if you compare a polygon theme with a point theme, ArcView finds points *inside* polygons and appends the attributes of the polygons to the points they contain. If you compare two point themes, ArcView finds the *nearest* point in the second theme to each point in the first theme. It then appends the attributes of points in the second theme to the corresponding points in the first theme.

Joining attributes based on containment

Suppose you're a wildlife biologist studying water sources on protected lands, such as national parks, national forests, and wildlife reserves. When water sources become scarce during the dry season, wild animals in these areas often migrate to grazing lands that support cattle. You'd like to anticipate this situation and take measures to prevent it. To do so, you'll need to know how many water sources each area has. You'll perform a spatial join to append the attributes of these areas to the water sources found within them. Then you'll perform a query to find the water holes that are inside protected areas and summarize their attributes to determine the number of water holes in each protected area.

Exercise 20a

1. If necessary, start ArcView. From the File menu, choose Open Exercise. In the Exercises scrolling list, select "ex20a," then click OK. When the project opens, you see a view with two themes, Water Holes and Range. Features in the Range theme are divided into three classes: Unprotected Areas, Protected Areas, and Unknown.

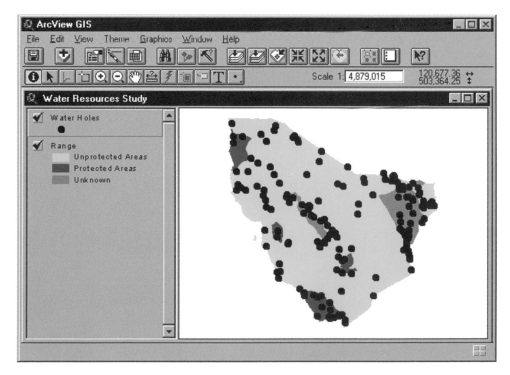

You want to know the characteristics of the area each water hole lies in, so you'll join the attributes of the Range theme to those of the Water Holes theme.

2. With the Water Holes theme active, hold down the Shift key and click on the Range theme to make both themes active. Then click the Open Theme Table button to open the attribute tables for both themes.

3. In the Attributes of Range table, click on the Shape field to make it active. Then do the same for the Attributes of Water Holes table. This field is common to both tables.

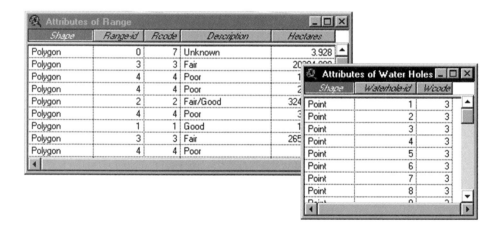

4. With the Attributes of Water Holes table active, click the Join button. ArcView joins the Attributes of Range theme table to the Attributes of Water Holes theme table based on the location of features in the two themes. Widen the table window so you can see the new fields that have been added.

Shape	Waterhole-id	Wcode	Range-id	Rcode	Description	Hectares
Point	1	3	1	1	Good	3984488.000
Point	2	3	1	1	Good	3984488.000
Point	3	3	6	6	Siblioi National Park	147378.000
Point	4	3	6	6	Siblioi National Park	147378.000
Point	5	3	1	1	Good	3984488.000
Point	6	3	1	1	Good	3984488.000
Point	7	3	1	1	Good	3984488.000
Point	8	3	1	1	Good	3984488.000

Attributes of Water Holes

Understanding spatial join. In a spatial join, you join two theme tables by using the Shape field as the common field. In an ordinary join, ArcView compares attribute values in the common field to match records. In a spatial join, it compares the locations of the features in each theme. If the destination table belongs to a point theme and the source table to a polygon theme, ArcView appends the attributes of each polygon to any and all points contained by that polygon. This is called an *inside* spatial join. If both the destination and source tables belong to point themes, ArcView appends the attributes of each point in the source table to the point nearest it in the destination table. This is called a *nearest* spatial join. Every spatial join is either inside or nearest, depending on the theme types represented by the source and destination tables. For more information, search for these Help Topics: *Spatial join, Performing spatial analysis with ArcView.*

The Attributes of Water Holes table (destination table) now has fields appended from the Attributes of Range table (source table). For each water hole, there is now a range code (Rcode) and description. Water holes with a range code of 6 are located in national parks, forests, and reserves. You'll build a query to select them.

5.   Make the view active. Both themes (Water Holes and Range) are active. Click on the Water Holes theme to make it the only active theme, then click the Query Builder button to display the Query Builder dialog box.

6. In the dialog box, double-click "[Rcode]" in the Fields list, then click the "=" button, then double-click "6" (national parks, forests, and reserves) in the Values list. Click New Set to select all the water holes in the national parks, forests, and reserves.

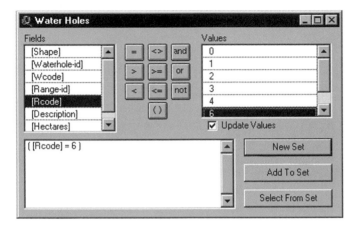

7. Close the Query Builder. You see the selected water holes highlighted in yellow in the view.

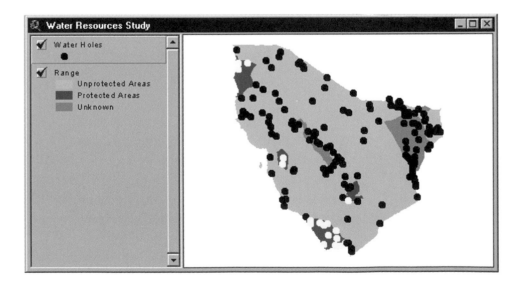

To see the attributes of the selected water holes, you'll open the theme table.

8. Click the Open Theme Table button, then click the Promote button to move the selected records to the top of the table.

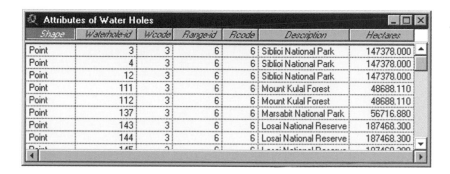

Next you'll summarize the selected water holes based on the values in the Description field to determine the number of water holes in each park, forest, and reserve.

9. In the Attributes of Water Holes table, make the Description field active. From the Field menu, choose Summarize. The Summary Table Definition dialog box displays.

10. Click the Save As button to navigate to the *drive:\directory* where you want to save the summary table ArcView creates and call it **watdes.dbf.** Click OK to create the summary table. (Because you are summarizing without using the Add button to create other fields, the summary table will contain only two fields: a Description field and a Count field.)

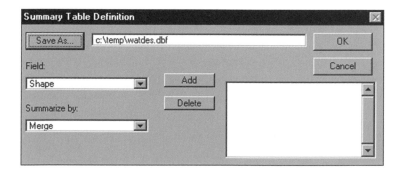

ArcView creates one record for each park, forest, and reserve that's named in the Description field. The Count field lists the number of water holes in each one.

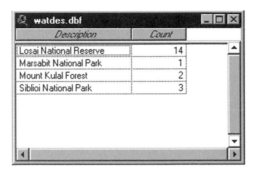

Now you know which protected areas have water sources and how many sources are in each area. You can study these sources to find out which can support local wildlife throughout the dry season, and which can't.

Joining attributes based on proximity

As a way of raising money to maintain protected areas, the government would like to sponsor camera safaris to water holes in those areas. Now your job is to evaluate access to water holes. You've discovered that only one of the four protected areas is accessible by land. The other three are too remote. As an alternative, the government would like to fly tourists into these areas. You'll need to know how far each water hole is from the nearest landing strip.

Because tourists will have to travel over bumpy, unimproved roads to get from landing strips to water holes, you want to find those water holes within 20 kilometers of a landing strip, to minimize driving time. To solve this problem, you'll perform a spatial join to append the attributes of landing strips to the water holes found nearest to them. ArcView calculates the distance from each water hole to the nearest landing strip and, in a field called "Distance," appends this information to each water hole. You'll perform a query on this field to find out which water holes are located less than 20 kilometers from a landing strip.

Exercise 20b

1. If *ex20a.apr* is open, close the Water Resources Study view and the open tables, then open the Water Holes in Protected Areas view. Otherwise, choose Open Exercise from the File menu. In the Exercises scrolling list, select "ex20b," then click OK. When the project opens, you see a view with three themes: Airports, Water Holes, and Range.

In the Range theme, the three remote protected areas are labeled. The Water Holes theme shows only the water holes inside these protected areas. The Airports theme (active) shows all the landing strips for the entire region.

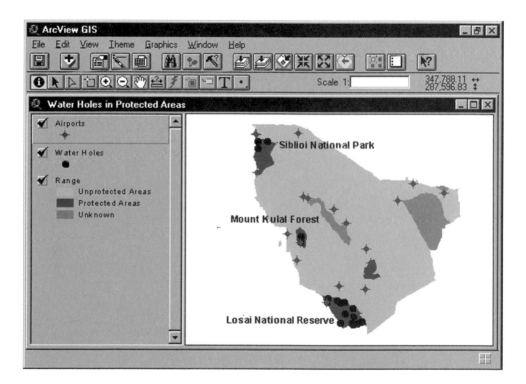

2. With the Airports theme active, hold down the Shift key and click on the Water Holes theme. Now both themes are active. Click the Open Theme Table button to open both attribute tables.

3. In the Attributes of Airports table (source table), make the Shape field active. Do the same for the Attributes of Water Holes table (destination table).

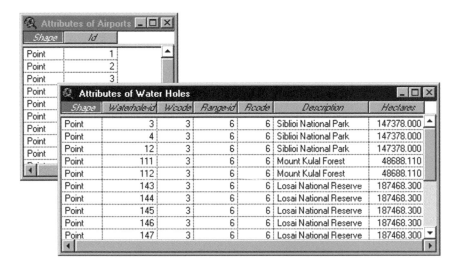

4. With the Attributes of Water Holes table active, click the Join button to join the Attributes of Airports table to the Attributes of Water Holes table. Widen the table so you can see the new fields that were added.

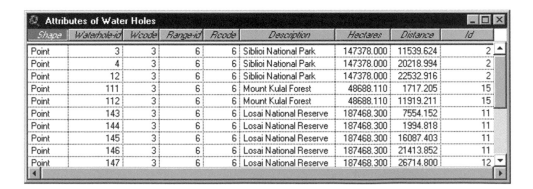

ArcView uses the *nearest* relationship to find the landing strip that each water hole lies closest to. The attributes of the landing strip (Id) are then appended to the attributes of the water hole. ArcView also calculates the distance between each water hole and the closest landing strip and places this value in a field called "Distance." (The distance values are calculated in the view's map units. For a review of map units, see chapter 11.)

Whenever ArcView uses the *nearest* relationship in a spatial join, a distance field is added to the joined table.

Now you'll build a query to find the water holes that are less than 20 kilometers from a landing strip.

5. With the Attributes of Water Holes table active, click the Query Builder button to display the Query Builder dialog box. In the box, double-click "[Distance]" in the Fields list, then click the "<" button, then type **20000** (20,000 meters is equal to 20 kilometers) in the query text box.

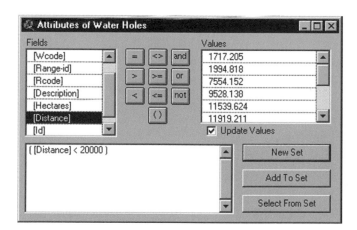

6. Click New Set to select all the water holes that are less than 20 kilometers from a landing strip. Close the Query Builder.

7. Click the Promote button to move the selected records to the top of the table.

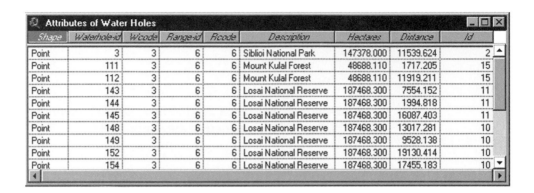

Shape	Waterhole-id	Wcode	Range-id	Rcode	Description	Hectares	Distance	Id
Point	3	3	6	6	Siblioi National Park	147378.000	11539.624	2
Point	111	3	6	6	Mount Kulal Forest	48688.110	1717.205	15
Point	112	3	6	6	Mount Kulal Forest	48688.110	11919.211	15
Point	143	3	6	6	Losai National Reserve	187468.300	7554.152	11
Point	144	3	6	6	Losai National Reserve	187468.300	1994.818	11
Point	145	3	6	6	Losai National Reserve	187468.300	16087.403	11
Point	148	3	6	6	Losai National Reserve	187468.300	13017.281	10
Point	149	3	6	6	Losai National Reserve	187468.300	9528.138	10
Point	152	3	6	6	Losai National Reserve	187468.300	19130.414	10
Point	154	3	6	6	Losai National Reserve	187468.300	17455.183	10

Now you know which water holes are closest to landing strips. These sites meet your criteria for good camera safari sites. However, you want to evaluate the remaining (unselected) sites for future development, so you'll change the selected set to water holes that are more than 20 kilometers away from landing strips.

  8. Click the Switch Selection button to select the water holes that are more than 20 kilometers away from a landing strip, then Promote the selected records.

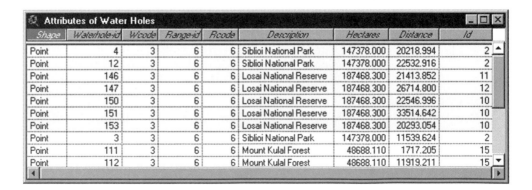

Shape	Waterhole-id	Wcode	Range-id	Rcode	Description	Hectares	Distance	Id
Point	4	3	6	6	Siblioi National Park	147378.000	20218.994	2
Point	12	3	6	6	Siblioi National Park	147378.000	22532.916	2
Point	146	3	6	6	Losai National Reserve	187468.300	21413.852	11
Point	147	3	6	6	Losai National Reserve	187468.300	26714.800	12
Point	150	3	6	6	Losai National Reserve	187468.300	22546.996	10
Point	151	3	6	6	Losai National Reserve	187468.300	33514.642	10
Point	153	3	6	6	Losai National Reserve	187468.300	20293.054	10
Point	3	3	6	6	Siblioi National Park	147378.000	11539.624	2
Point	111	3	6	6	Mount Kulal Forest	48688.110	1717.205	15
Point	112	3	6	6	Mount Kulal Forest	48688.110	11919.211	15

The highlighted records are water holes that are more than 20 kilometers from a landing strip. To examine the locations of these water holes, you'll look at the view.

9.   Make the view active. Water holes more than 20 kilometers from landing strips are highlighted.

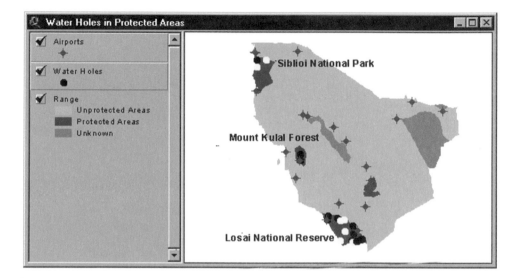

These water holes can be developed for future safaris by adding new landing strips or by improving the roads leading from them to the nearest existing landing strip.

ArcView helped you meet your initial objective of evaluating access to water holes for the purpose of offering camera safaris. You know how far each water hole is from the nearest landing strip, which ones are within 20 kilometers of existing landing strips, and which ones aren't. If the camera safari project is successful in raising revenues for protected areas, then both the local economies and the protected areas will benefit.

If you want to go on to the next chapter, leave ArcView running. Otherwise, choose Exit from the File menu.

SECTION 6

Presenting information

The next two chapters show you how to present information as charts and map layouts. In chapter 21, you'll chart an attribute in the theme table, then change the way the chart looks. You'll also query the chart and change the information it presents. In chapter 22, you'll create a map layout and add and manipulate all the elements you need to make it presentation-quality.

SECTION 6:
Presenting information

Working with charts

Creating charts

Modifying charts

Querying and editing charts

Working with charts

Just as views are excellent for presenting spatial information, charts are ideal for displaying tabular information. A chart references tabular data in an existing table. Charts enhance your presentation by providing a graphic representation of the attributes associated with map features. With a chart, you can turn a list of complicated figures into brightly colored graphics that clarify complex relationships at a glance.

In ArcView GIS, charts are simple to create and change. In this chapter, you'll create a chart, modify its characteristics, and use it to access information from its source table.

Creating charts

Suppose you're the marketing director for Miles From Nowhere, Inc., a company that offers recreational wilderness trips to remote parts of the world. Your company already offers treks across the Australian outback, cross-country skiing in Greenland, African safaris, and Amazon jungle bushwhacks. It's your job to come up with a new angle. Thumbing through an atlas one afternoon, you get inspired: what could be farther off the beaten path than an expedition to Siberia?

Your preliminary research reveals that population density throughout Siberia is less than three people per square mile, summer temperatures average between 50 and 70 degrees Fahrenheit, and rails run east and west, linking a number of cities (Krasnoyarsk, Bratsk, Irkutsk, Ust-Kut) that could serve as starting points for journeys into the interior. There are several major navigable waterways, suggesting that a river trip may be a good way to travel.

You'll use ArcView to develop your plan and present it at the quarterly business meeting. In your presentation, you'll include a view of Siberia and its major rivers as well as a chart comparing the lengths of the rivers.

Exercise 21a

1. If necessary, start ArcView. From the File menu, choose Open Exercise. In the Exercises scrolling list, select "ex21a," then click OK. When the project opens, you see a view with two themes, Siberian Rivers and Far East (a theme showing the region of Siberia and adjacent countries). You also see the attribute table for the Siberian Rivers theme.

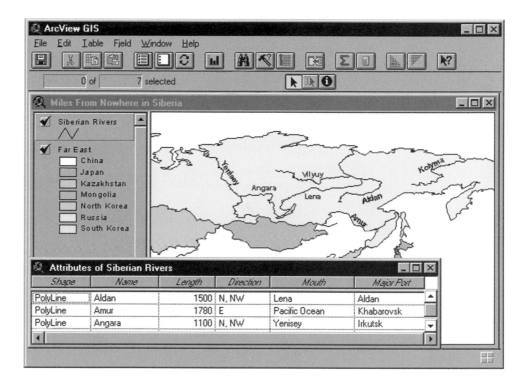

You want to create a chart comparing the values of the Length attribute in the table. By default, all records in the table will be charted. If there is a selected set of records, only the selected set will be charted. The chart, the table, and the theme in the view are all dynamically linked, which means that changes to any one of these are immediately reflected in the others. (You can also import and chart a table that's not associated with any theme. In that case, the table and the chart are linked to each other, but neither is linked to the view.)

2. Click the Create Chart button to open the Chart Properties dialog box.

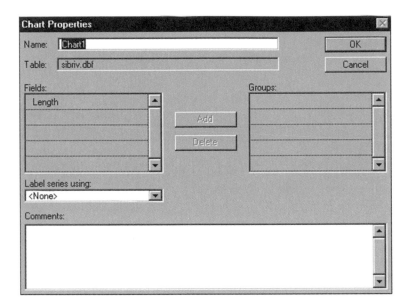

3. In the Name input box, change the name to **River Lengths.**

The Fields scrolling list displays the names of all chartable fields in the active table. In this case, there is only one. (Hidden and non-numeric fields can't be charted.)

4. Click on "Length" in the Fields scrolling list, then click the Add button. This puts the Length field in the Groups list to be charted.

Below the Fields list, a drop-down list (Label series using) displays the fields you can use to label each item in your chart.

ENVIRONMENTAL SYSTEMS RESEARCH INSTITUTE, INC.

5. In the lower drop-down list (Label series using), select "Name."

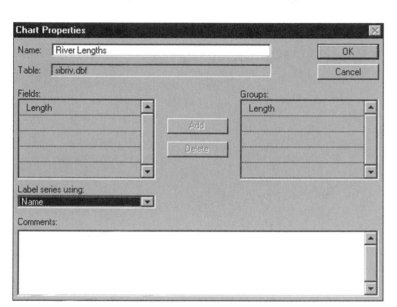

6. Click OK. ArcView plots the river lengths in the default chart format.

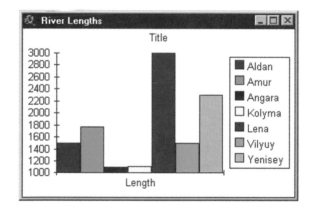

A colored bar, or *data marker,* shows the length of each river. The name of each river appears in the chart's legend. The y-axis is a scale showing the range of values for the river lengths.

Choosing a chart format. ArcView offers six types of charts: area, bar, column (the default), line, pie, and x,y scatter. The type you choose depends on the nature of your data and the message you want to convey. Line charts, for instance, are good for showing changes in values over time. Pie charts, on the other hand, show relationships between parts and the whole. For more information, search for this Help Topic: *Choosing a chart format.*

The River Lengths chart compares the values of a single variable (length). For this type of comparison, a column or bar chart is best.

Modifying charts

When you create a chart, ArcView displays it with the default format, axes, title, and legend. ArcView also applies a default set of colors to the data markers. You can change a chart's appearance by modifying any of these characteristics.

You want to make the chart more presentable for your business meeting by adding a title and axis labels. You also want to change the units on the y-axis.

Exercise 21b

1. If *ex21a.apr* is open, continue. Otherwise, choose Open Exercise from the File menu. In the Exercises scrolling list, select "ex21b," then click OK. When the project opens, you see a view and the active River Lengths chart. (If you have *ex21a.apr* open, the table will still be visible as well.)

ENVIRONMENTAL SYSTEMS RESEARCH INSTITUTE, INC.

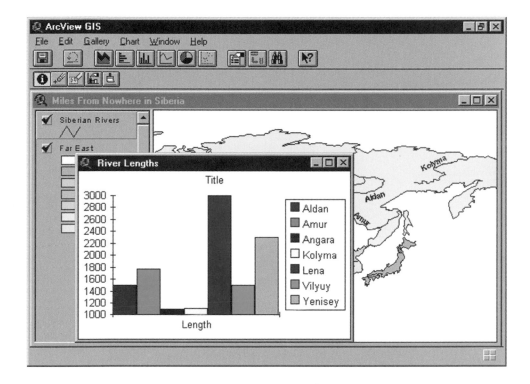

Next you'll refine the chart. (You may want to enlarge the chart window so the elements are easier to see.)

Notice the values on the y-axis. The lowest is 1,000, the highest 3,000, and the increment 200. The chart would be more meaningful if the scale started at zero.

2. With the chart active, click on the Chart Element Properties tool, then click anywhere on the y-axis of the chart. The Chart Axis Properties dialog box displays.

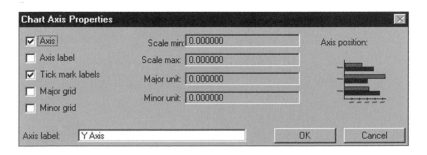

3. In the Scale max text box, highlight the default value and type **3500.** This sets the upper limit on the y-axis. (The lower limit is already set to 0, although this new value hasn't yet been applied to the chart.)

4. In the Major unit text box, replace the default value with **500** to adjust the increment along the y-axis.

5. In the Axis label text box at the bottom of the dialog box, highlight "Y Axis" and replace it with the word **Miles.**

6. Click on the Axis label check box (along the left side of the dialog box) to turn it on. The Chart Axis Properties dialog box now looks like this:

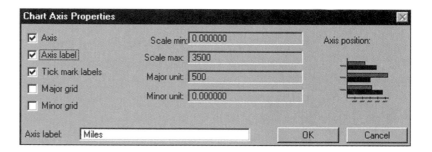

7. Click OK. The chart redraws with the changes you made.

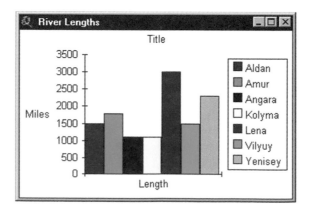

The chart looks better, but it's still difficult to follow the y-axis labels across to the data markers.

8. Click again on the chart's y-axis to reopen the Chart Axis Properties dialog box. Click on the Major grid check box (along the left side of the dialog box), then click OK. The chart redraws with gridlines that make it easier to read.

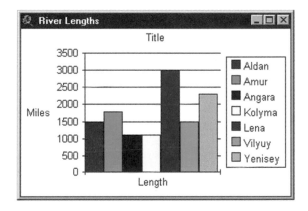

Notice that the attribute you're charting, "Length," is currently displayed along the bottom of the chart. This is the Group label. You'll also add an x-axis label to the bottom of the chart.

9. Click on the chart's x-axis to open its Chart Axis Properties dialog box.

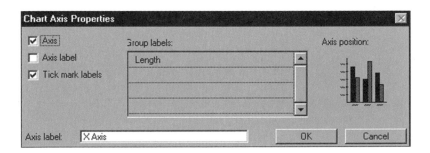

10. In the Axis label text box at the bottom of the dialog box, highlight "X Axis" and replace it with the words **Note: The Mississippi River is 2470 miles long.** Click on the Axis label check box (along the left side of the dialog box) to turn it on, then click OK. Your chart looks like this:

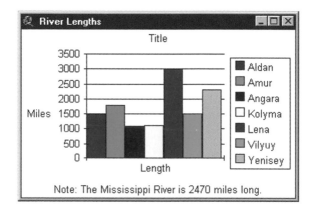

Changing the chart's legend. You can edit the labels that appear in the legend or change the position of the legend using Chart Legend Properties. Access the Chart Legend Properties dialog box by clicking on the chart's legend with the Chart Element Properties tool. For more information, search for these Help Topics: *Chart Legend Properties (Dialog box), Chart Element Properties tool.*

Now you'll give your chart a title.

11. Click on the word "Title" at the top of the chart to open the Chart Title Properties dialog box.

The title has five fixed positions you can click on: top, bottom, left, right, or middle. (If you choose the middle position, you can place the title anywhere you want.) You'll leave the title at the top (the default position).

12. In the text box, type the words **Major Siberian Rivers,** then click OK. The chart redraws with the new title.

Your chart looks good, but you want to change the gray color representing the Yenisey River to something more vivid. For this you'll use the Chart Color tool.

13. Click on the Chart Color tool in the Chart tool bar. When the Symbol Palette displays, choose the Color Palette.

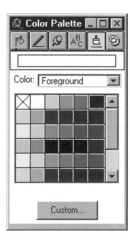

You can use the Chart Color tool to change the color of any chart element, such as data markers, title, and labels.

14. In the Color Palette, click on the "russet" square (the one diagonal to black). A black border appears around it to indicate that it's selected. Now bring the cursor over to the chart and click on the gray data marker for the Yenisey River. The color of the data marker changes to russet, and the legend symbol changes to match. Close the Color Palette.

Your chart now looks like this:

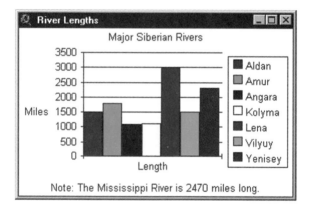

If you want to go on to the next exercise, leave the project open.

Querying and editing charts

Because ArcView charts are dynamically linked to the tables from which they're created, you can get information from the table simply by clicking on the chart. You can edit or select records in the chart's source table and see the change reflected immediately in the chart. You can also modify the chart by deleting data markers.

Your chart is almost ready for presentation at the quarterly business meeting. You realize that attendees will have questions about the information behind the chart, and you must be prepared. You also want to remove any information from the chart that's not useful to your presentation.

Exercise 21c

1. If either *ex21a.apr* or *ex21b.apr* is open, continue. Otherwise, choose Open Exercise from the File menu. In the Exercises scrolling list, select "ex21c," then click OK. When the project opens, you see the view and the active chart.

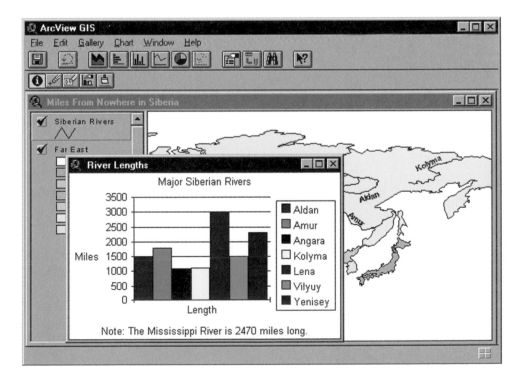

2. From the Window menu, select Attributes of Siberian Rivers to make the attribute table visible. Move and resize the document windows as necessary so that you can see the chart, the table, and the upper part of the view.

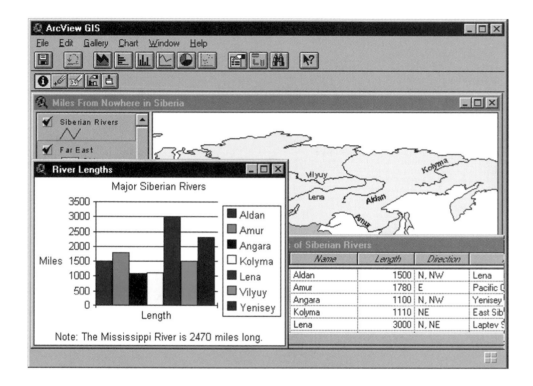

Each data marker in your chart corresponds to a record in the attribute table and a feature in the view. Using the Identify tool, you can click on any data marker to get information from the table about that marker. The corresponding record in the table is briefly highlighted in black and the feature flashes in the view.

3. Make sure the chart is active by clicking on its title bar. Choose the Identify tool from the Chart tool bar if it's not already selected. Now click on the yellow data marker for the Kolyma River. The corresponding tabular data displays in the Identify Results dialog box. The Kolyma River flashes in the upper right corner of the view and the record for Kolyma highlights briefly in the table.

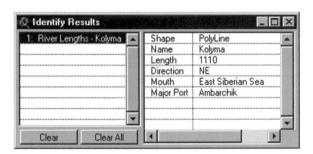

Clicking on additional data markers adds them to the Identify Results dialog box.

4. Close the Identify Results dialog box when you're finished.

Looking at the view, you can see that the Kolyma River in northeastern Siberia is extremely remote, making it too expensive and difficult to launch an expedition there. You want to remove the Kolyma from your chart.

5. If the chart isn't already active, make it active. Click on the Erase tool and click on the yellow data marker for the Kolyma River.

The Kolyma data marker disappears from the chart. Notice also that the colors of the data markers have shifted one position to the right, eliminating the russet color.

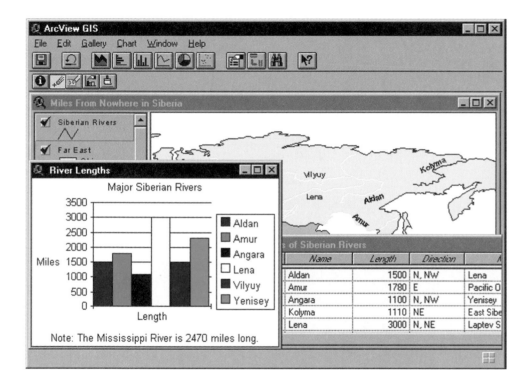

Note that all records in the table and all rivers in the view, except Kolyma, are selected. When you delete a data marker from the chart, the record corresponding to that data marker is unselected in the table and all other records are selected. Only the selected records are charted.

Adding and deleting data markers. You can add data markers to a chart by selecting more records in the table, or delete data markers by unselecting records or by using the Erase tool. If the table is an attribute table for a theme, you can add or remove data markers in the chart by selecting or unselecting features in the view. For more information, search for these Help Topics: *Adding and erasing data markers on a chart, Erase tool.*

You're ready to present your ideas at the quarterly business meeting. The chart, together with the view, communicates something of the vastness of Siberia, focusing attention on its rivers as perhaps the best means of exploring the interior. You realize that further research will be needed to refine your plan. You'll need information on river navigability, the business climate in Russia, cost analyses, and so on.

If you want to go on to the next chapter, leave ArcView running. Otherwise, choose Exit from the File menu.

SECTION 6:
Presenting information

Creating
map layouts

Making a basic map layout

Adding charts and tables

Adding the finishing touches and printing

Creating map layouts

You've seen how ArcView GIS creates views, tables, and charts. Each of these documents presents information in a different format. But what if you want to display all of these formats at the same time on your screen, or print them on a single piece of paper? And what if you want to add a scale bar, north arrow, border, and title to create a presentation-quality map?

You can do all these things with ArcView. It allows you to dynamically place views, tables, charts, images, and any graphic elements you want in one document, called a *layout*. You can think of a layout as representing the final piece of paper your map will be printed on. If you change your mind later, you can add, remove, resize, and move each element in a layout as required. Instead of a static map, you have a dynamic document that adapts to your needs.

Making a basic map layout

Suppose you're publishing a book on the social and economic growth of Canada. You want to include a population density map to show where most Canadians live and work. You'll create a layout that includes a view of population density by province, a table of population statistics, and a chart comparing the population of major Canadian cities.

Exercise 22a

1. If necessary, start ArcView. From the File menu, choose Open Exercise. In the Exercises scrolling list, select "ex22a," then click OK. When the project opens, you see a view, a table, and a chart.

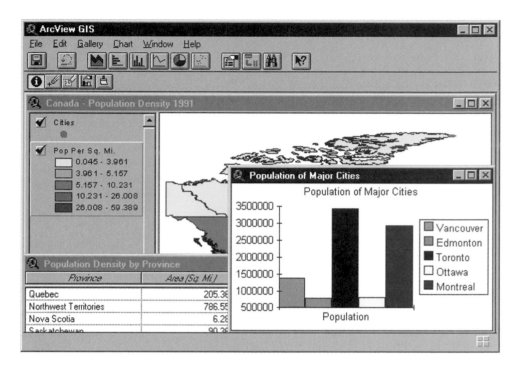

You'll create a layout containing these documents.

2.   From the Window menu, select *ex22a.apr* to open the Project window. Click on the Layouts icon in the Project window, then click New. A blank layout page appears.

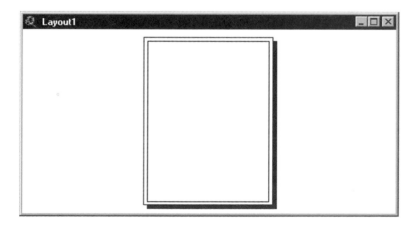

> **Note: If you're on the UNIX operating system, the default layout page is landscape (horizontal) instead of portrait (vertical).**

3. Choose Page Setup from the Layout menu. Click the landscape (horizontal) button in the Orientation field.

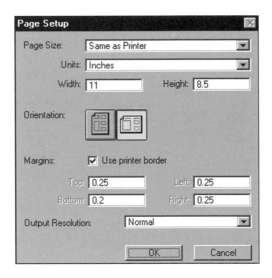

4. Click OK. The layout changes orientation to landscape.

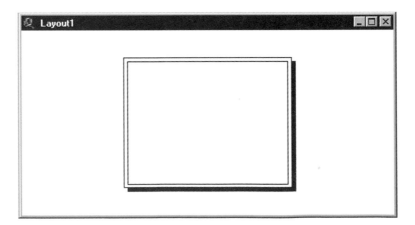

The layout page is too small to work with effectively, so you'll enlarge it.

5.   Click on the icon in the left corner of the layout title bar and select Maximize.

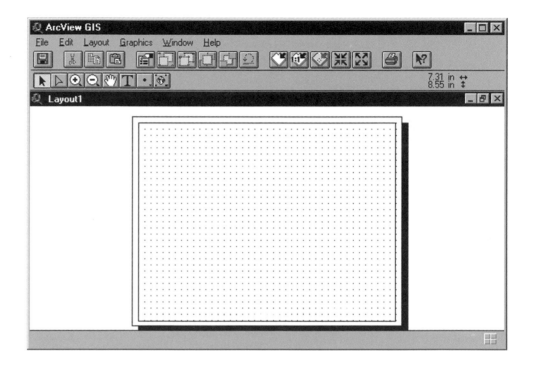

Now that the layout page is larger, you see grid dots. These dots are used to snap elements to precise locations in the layout.

Grid dots appear only on the screen, not on the printout. If you don't want to show the grid as part of an on-screen presentation, you can hide it by clicking Hide Grid in the Layout menu.

Before you add elements to your layout, you'll give the layout a new name.

6. From the Layout menu, select Properties. Click in the Name field and change the name to **Population Density 1991.** Click OK.

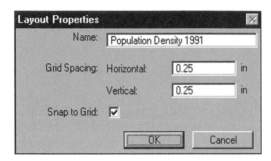

Now that you've finished the basic page setup for the layout, you can begin to add views, tables, charts, and other elements to the layout using the Frame tool.

> **Adding documents to a layout.** Each document you add to a layout has its own box, or *frame.* There's a different tool for creating each document frame, and each has its own properties. For example, you can use the Chart Frame tool to draw a frame for your population chart of Canadian cities. After you draw the frame, the Chart Frame Properties dialog box displays, which enables you to choose the chart you want from all of the available charts. After specifying the properties you want, the chart draws inside the frame. For more information, search for these Help Topics: *Frame tool, Types of frames.*

The Frame tool is a set of drop-down tools on the Layout tool bar. It includes an icon for each of the following types of frames: view, legend, scale bar, north arrow, chart, table, and picture. You'll use the View Frame tool first.

7. Click on the Frame tool and hold down the mouse button to display the drop-down tools. Select the View Frame tool.

8. Click in the upper left corner of the layout page and hold down the mouse button as you drag a box for the view frame. (The exact size of the frame you draw doesn't matter; you can fix it later.) When you release the button, the View Frame Properties dialog box displays.

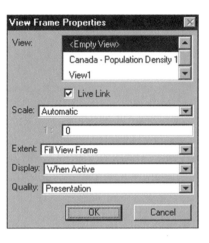

9. Click on "Canada - Population Density 1991" in the View scrolling list.

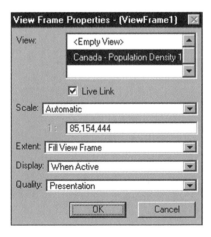

Live Link is checked by default, specifying that the view frame will be linked to the view. This means that changes to the view are automatically reflected in the view frame. The Scale field is set to "Automatic," so ArcView calculates the appropriate scale based on the size of the view frame you drew. The Extent field determines what will happen when the view frame dimensions aren't proportional to the view document dimensions. If Extent is set to "Fill View Frame," you may see more data in the view frame than you see in the view. If Extent is set to "Clip to View," the view frame will only show the data in the view, even if a portion of the view frame is empty.

The Display drop-down list is set to "When Active," meaning that the view will display in the view frame only when the view document is open in the project. When the view document is closed, the view frame will display a gray box. You can reset Display to "Always," so that the view always appears in the frame, even when the view document is closed. The Quality field can be set to "Draft" if you don't want to wait for the view frame to redraw (it will be a gray box), or to "Presentation" if you want to see the view frame contents.

10. Click OK. The Canada - Population Density 1991 view draws in the view frame. The four black handles indicate that this is the currently selected frame.

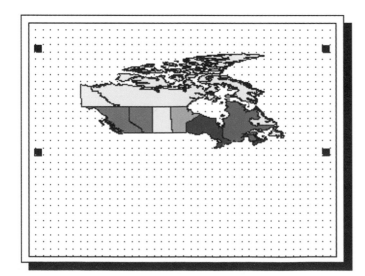

ENVIRONMENTAL SYSTEMS RESEARCH INSTITUTE, INC.

You're ready to add a legend to your layout. The legend represents the view's Table of Contents and stays inside a legend frame in the layout.

11. Click on the Frame tool and hold down the mouse button to display the drop-down tools. Select the Legend Frame tool.

12. Below the view frame, drag a box for the legend frame. The Legend Frame Properties dialog box displays.

The View Frame scrolling list shows the view frames that have already been placed in the layout.

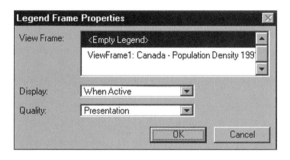

13. Click on "View Frame 1: Canada - Population Density 1991." Then click OK. The legend associated with this view frame draws on the layout.

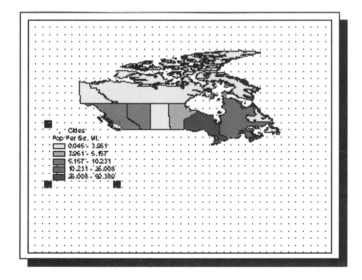

If you don't like the size of the legend, you can change it with the Pointer tool.

14. Click on the Pointer tool and move the cursor over one of the four handles around the legend frame until it changes to a double-headed arrow. Hold the mouse button down and drag the handle to change the size. The legend frame snaps to the nearest grid point and the legend redraws.

To move a layout element without resizing it, select it with the Pointer tool, move the cursor over it (not on a handle) until the cursor changes to a four-headed arrow, then click and drag it to a new location. You can make very small adjustments by using the arrow keys to move the element in small fractions of the grid spacing.

Next you'll add a scale bar to the layout. The scale bar has its own frame in the layout and is associated with a view frame.

15. Choose the Scale Bar Frame tool from the drop-down tools. Drag a scale bar frame box in the layout just as you did the view and legend frames. The Scale Bar Properties dialog box displays.

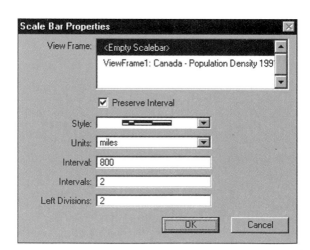

ENVIRONMENTAL SYSTEMS RESEARCH INSTITUTE, INC.

16. Select "View Frame 1: Canada - Population Density 1991." Then click on the drop-down arrow for the Style field and choose a scale bar style. Click OK. The scale bar draws inside the scale bar frame in the layout.

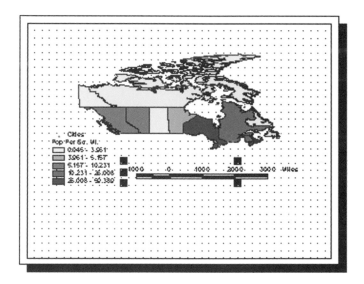

Remember, if you don't like the size or location of any layout element, you can move or resize its frame using the Pointer tool.

Understanding scale bars. The scale bar draws to the size that accurately represents the scale of the view frame. When the view frame is live-linked to the view, the scale bar updates to reflect any change to the scale of the view. For more information, search for these Help Topics: *Adding a scale bar to a layout, Scale bar frame tool.*

Finally, you'll add a north arrow to the layout.

17. Select the North Arrow Frame tool from the drop-down tools and drag a north arrow frame box in the layout. The North Arrow Manager dialog box displays.

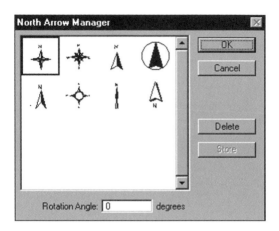

18. Select a north arrow and click OK. The north arrow draws inside its frame.

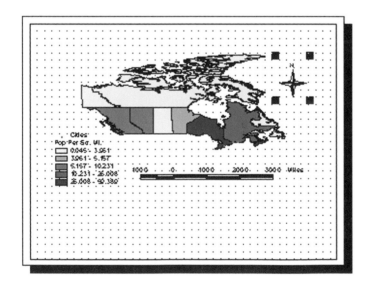

If you want to go on to the next exercise, leave the project open.

Adding charts and tables

The map layout for your book on the social and economic growth of Canada is almost finished. It includes a view showing population density by province with a legend, scale bar, and north arrow. To enhance your presentation, you'll add a chart comparing the population of major cities in Canada and a table of population statistics.

Exercise 22b

1. If *ex22a.apr* is open, continue. Otherwise, choose Open Exercise from the File menu. In the Exercises scrolling list, select "ex22b," then click OK.

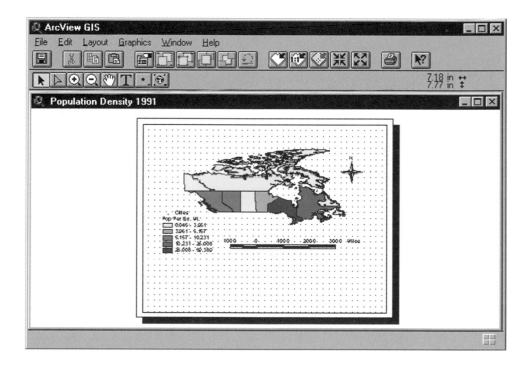

2. Select the Chart Frame tool from the drop-down tools and drag a chart frame in your layout. The Chart Frame Properties dialog box displays.

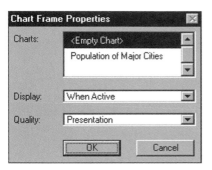

3. Select "Population of Major Cities" and click OK. The chart draws inside the chart frame.

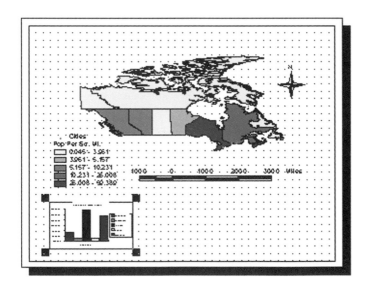

4. Next, select the Table Frame tool from the drop-down tools and drag a table frame in your layout. The Table Frame Properties dialog box displays.

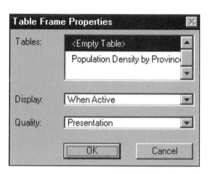

5. Choose the "Population Density by Province" table and click OK. The table draws in your layout.

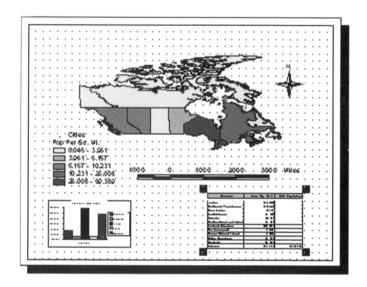

Importing scanned images. Want to really add impact to your layout? The Picture Frame tool lets you add images to your layout. You can import scanned images in several different file formats into a picture frame. Just choose the Picture Frame tool and create a picture frame. Then in the Picture Frame Properties dialog box, specify the file to import into the picture frame. For more information, search for these Help Topics: *Picture frame tool, Adding imported graphics to a layout.*

If you want to go on to the next exercise, leave the project open.

Adding the finishing touches and printing

Congratulations! You've added all the frames you need for your final layout presentation called "Population Density 1991." Now it's time to add some finishing touches (a title and a neatline) to make your layout look professional.

Exercise 22c

1. If *ex22b.apr* is open, continue. Otherwise, choose Open Exercise from the File menu. In the Exercises scrolling list, select "ex22c," then click OK.

ENVIRONMENTAL SYSTEMS RESEARCH INSTITUTE, INC.

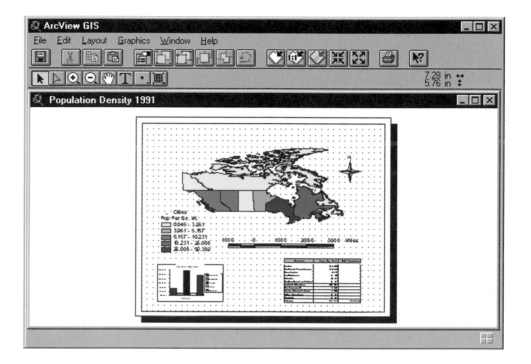

To give your layout a title, you'll use the Text tool.

 2. Click on the Text tool in the Layout tool bar, then click inside the layout, near the top, where you want your title to start. The Text Properties dialog box displays.

3. Type in **Population Density 1991,** then click OK.

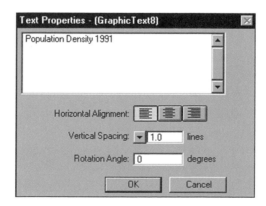

The title draws in the layout.

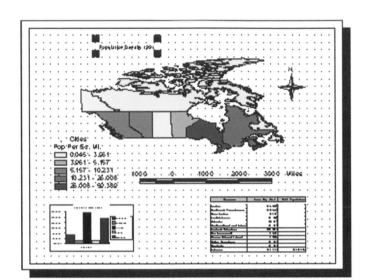

The title is too small. You can make it bigger by using the Symbol Window to change the font size.

4. Click on the Window menu and select Show Symbol Window. The Symbol Window displays.

5. Select the Font Palette if it's not already selected.

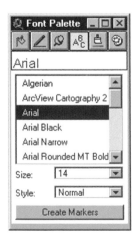

6.   Click on the Size drop-down arrow and select "36." The title changes to 36-point text. Close the Font Palette.

If your title isn't centered, use the Pointer tool to drag the text to the desired position.

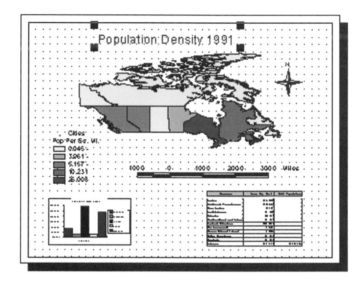

Aligning frames and graphics. You can align the frames and graphics in your layout with one another. Use the Pointer tool to select the elements to align, then choose Align in the Graphics menu. In the Align dialog box, use the buttons to align selected elements along the top, bottom, right, left, or center of your layout. For more information, search for these Help Topics: *Align (Dialog box), Aligning graphics.*

Next you'll add a neatline.

7. Select the Rectangle tool from the Draw tool drop-down choices. Drag a rectangle around your layout so it surrounds all the frames and graphics.

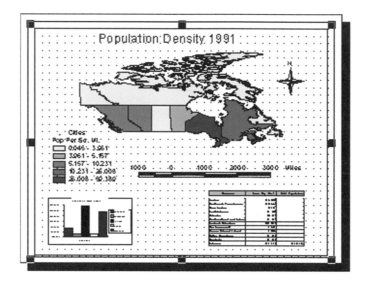

If all the elements in your layout don't fit inside the neatline, you may need to rearrange them. With the rectangle selected, choose Send to Back from the Graphics menu to place the rectangle behind the other elements in your layout. Then use the Pointer tool to select and move any elements.

Now that you've completed the layout page for your book, you're ready to print it.

ENVIRONMENTAL SYSTEMS RESEARCH INSTITUTE, INC.

8. From the File pulldown menu, select Print. The Print dialog box displays. (If you have a PostScript® printer, you'll see additional options for PostScript in the Print dialog box.) If you have a printer connected to your computer and it's turned on, click OK to print the layout. Otherwise, click Cancel.

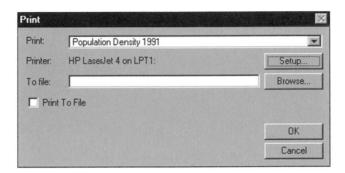

> **Storing a layout as a template.** You've done a lot of work to create this layout and position all the frames exactly where you want them. Suppose you have many layouts to create and you want each of them to have the same look. Because the current layout can be stored as a template for future layouts, you won't have to do all the work over again. For more information, search for these Help Topics: *Template Properties (Dialog box), Using a layout as a cartographic template.*

You've seen how to display a view, table, chart, and other graphic elements in a layout. Because your layout is a dynamic document, you can easily change it anytime you need to.

If you want to go on to the next chapter, leave ArcView running. Otherwise, choose Exit from the File menu.

SECTION 7

Creating your own data

In the next four chapters, you'll learn several ways to create your own data. In chapter 23, you'll convert features from existing themes to create new themes. You'll also create a new theme by drawing shapes and adding attributes. In chapter 24, you'll modify shapes by editing them, merging them, and splitting them. You'll define how the attribute table is affected when you merge and split shapes. In chapter 25, you'll create a point theme from a file of x,y coordinates. In chapter 26, you'll create a point theme from a list of addresses by matching the list against a theme containing address information.

SECTION 7:
Creating your own data

Creating
shapefiles

Creating shapes from features

Drawing polygon shapes

Drawing point and line shapes

CHAPTER 23

Creating shapefiles

In chapter 8, you learned that ArcView GIS has its own file format, called a shapefile. Features saved in ArcView's shapefile format are called shapes. There are advantages to using shapefiles over other file types: shapes are the only kind of feature you can edit in ArcView, and themes created from shapefiles draw faster than themes created from other file types.

In this chapter, you'll create shapefiles in two ways: by converting existing features into shapes and by drawing new shapes.

First, you'll convert selected features from an existing theme into shapes and save them in a shapefile. You can convert features in themes created from ARC/INFO coverages, SDE database files, and CAD drawings into shapes. You can also "convert" features that are already shapes and save them in a new shapefile. This lets you extract selected features from a large theme and create a smaller theme with only the desired features.

Second, you'll create a new theme, draw new shapes in the theme, and save it as a shapefile. In the next chapter, you'll learn how to edit existing shapes.

ENVIRONMENTAL SYSTEMS RESEARCH INSTITUTE, INC.

Creating shapes from features

Imagine that you're a geologist studying volcanic activity in the Mojave Desert. You've collected several rock samples and located the collection sites on an air photo. Now you want to use the air photo to create a feature-based map of the area.

Exercise 23a

1. If necessary, start ArcView. From the File menu, choose Open Exercise. In the Exercises scrolling list, select "ex23a," then click OK. When the project opens, you see an air photo image of the volcanic field you're studying. Two cinder cones and a dark lava flow issuing from the lower cone are visible. The Sample locations theme shows the sites where you gathered specimens.

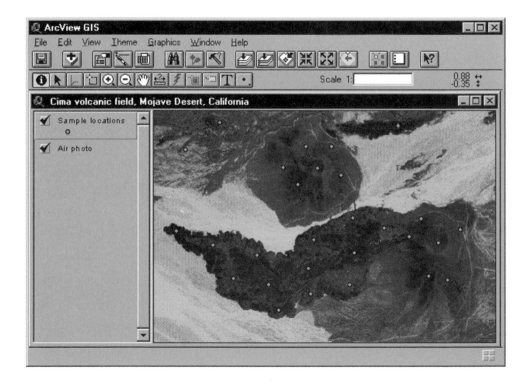

You plan to do a detailed study of the cinder cone in the lower half of the view. You want to create a theme that contains only the sample locations in your study area. The Sample locations theme is already stored in shapefile format, but because many sample points are outside your study area, you'll create a new shapefile to store only the points you want in it.

You can isolate the sample locations you want to save as shapes by selecting them with a graphic.

2. Make sure the Sample locations theme is active. Click and hold the Draw tool to display the drop-down list of tools. Select the Rectangle tool.

3. Draw a rectangle that includes the sample locations shown. (Your rectangle doesn't have to exactly match the one below. If you don't like the rectangle you draw, you can press the Delete key and draw a new one.)

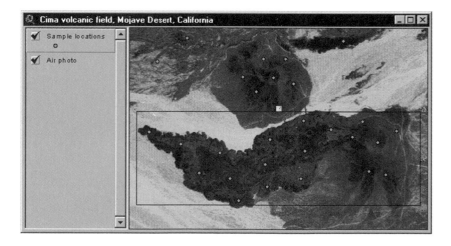

The rectangle you just drew is a graphic, not a shape. A graphic is a geometric object that is drawn on top of a view and is not associated with any theme. A graphic is not a map feature and has no attributes. A shape, by contrast, is a map feature associated with a particular theme, and it has attributes in a theme table.

4. Click the Select Features Using Graphic button. The features of the Sample locations theme that lie within the rectangle are selected and highlighted in yellow.

Now you'll convert the selected points to shapes in a shapefile, and add the shapefile as a theme in the view.

5. From the Theme menu, select Convert to Shapefile. The Convert Sample locations dialog box displays.

6. Select a *drive:\directory* where you want to save the shapefile. In the File Name text box, change the name to **samples.shp,** then click OK.

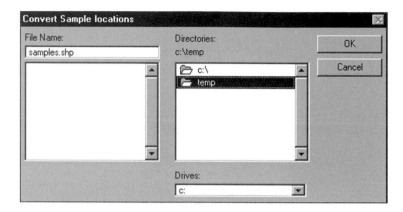

The Convert to Shapefile dialog box displays with the question "Add shapefile as theme to the view?" If you answer Yes, the selected point features are saved as shapes in the samples.shp shapefile, and added as a theme to the view. If you answer No, the shapefile is created, but isn't added as a theme.

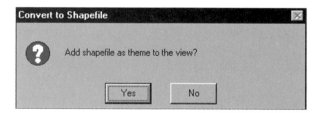

7. Click Yes. ArcView creates a new shapefile called samples.shp in your directory and adds the Samples.shp theme to the view.

8. Turn on the Samples.shp theme by clicking on its check box in the Table of Contents.

ArcView chooses a random color to display a new theme. The color shown here probably won't match yours. You can use the Legend Editor and Symbol Window to change the color and marker symbol (refer to chapter 10 for details).

You don't need the original Sample locations theme or the rectangle graphic any longer, so you'll delete them to simplify the view.

9. Make sure the rectangle is selected. (If it's not, click on it with the Pointer tool. Handles will appear around it when it's selected.) Press the Delete key.

10. Make sure that only the Sample locations theme is active. From the Edit menu, select Delete Themes. The Delete Themes dialog box displays and asks if you're sure you want to delete the theme.

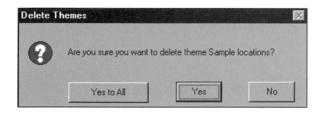

11. Click Yes. The Sample locations theme is removed from the project. (Don't worry: no files are deleted from the hard disk.) The view now looks like this:

You've created a new theme by selecting features in an existing theme, converting them to shapes in a shapefile, and displaying the new shapefile as a theme in the view. (If you convert a theme to a shapefile without selecting any features, *all* features in the theme will be converted.)

You've met your first objective. You now have a theme containing only those sample locations that are within the cinder cone and lava flow you're mapping.

Drawing polygon shapes

Your next task is to draw the outline of the cinder cone and the lava flow for your map. You'll create a new theme, draw two polygon shapes, then save the shapes in a shapefile.

Exercise 23b

1. If *ex23a.apr* is open, continue. Otherwise, choose Open Exercise from the File menu. In the Exercises scrolling list, select "ex23b," then click OK. When the project opens, you see a view displaying the air photo image and the sample sites for the study area.

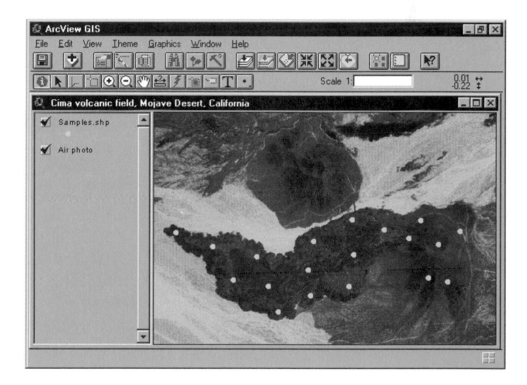

You may want to enlarge the ArcView window and the View window for this exercise.

To draw the cinder cone and lava flow, you'll create a new, empty theme, then add shapes to it.

2.   From the View menu, choose New Theme. The New Theme dialog box displays.

3.   Click on the Feature type drop-down arrow and select "Polygon," then click OK.

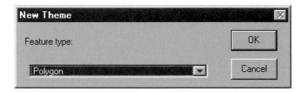

Another dialog box displays, asking you to name the theme.

4.   Specify the *drive:\directory* where you want to save the new theme, then change the file name to **volcano.shp.** Click OK.

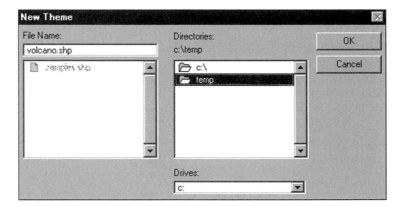

A new theme called Volcano.shp is added to the view. At the moment, it contains no features, which means that it contains no shapes. (Remember, a shape is just a map feature in shapefile format.)

Notice that the check box for the Volcano.shp theme has a dashed line around it, indicating that the theme can be edited.

How shapefiles are stored. A shapefile is actually stored as at least three files on your hard disk. The files have the format *filename*.shp, *filename*.shx, and *filename*.dbf. If the shapefile's theme table has been linked (see chapter 16), there will be files with .ain and .aih extensions. If the shapefile theme has been spatially joined (see chapter 20), or if you've used it in theme-on-theme selection (see chapters 17 through 19), there will be .sbn and .sbx files. All these files together make up the shapefile and are needed to display the shapefile as a theme. (When you move or copy a shapefile to a new location on your hard disk, you need to copy all the constituent files.) You can change a theme's name at any time by choosing Properties from the Theme menu. ArcView will keep track of the new theme name and all the associated files. For more information, search for this Help Topic: *ArcView shapefiles.*

First, you'll draw a polygon shape in the empty Volcano.shp theme to define the cinder cone.

5. From the Draw tool drop-down list, choose the Polygon tool.

6. Draw the boundary of the cinder cone as shown below. Click to start each side of the polygon, then double-click to finish the polygon. (The shape you draw doesn't need to look exactly like the one you see below. If you don't like the shape you draw, you can press the Delete key and start over.)

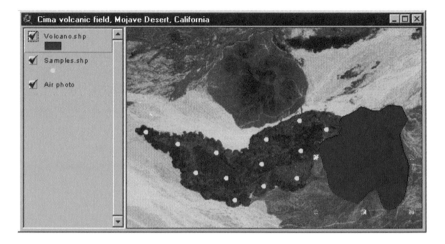

When you finish drawing the shape, selection handles appear around it.

> **Moving and resizing shapes.** When a theme is editable, you can move or resize a selected shape with the Pointer tool. If the shape is not already selected, click on it with the Pointer tool. (You can select multiple shapes by holding down the Shift key as you click.) To resize a selected shape, drag one of its selection handles with the Pointer tool. To move a selected shape, click on it with the Pointer tool and drag it to a new location. To unselect a shape, click somewhere else in the view. For more information, search for these Help Topics: *Editing line themes, Editing polygon themes.*

Now you'll switch drawing tools. You'll use the Autocomplete tool to form a common boundary between the cinder cone shape and the lava flow shape you're about to draw.

Using the Autocomplete tool

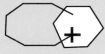

Click inside an
existing shape with
the Autocomplete
tool.

Draw a line that
forms a polygon
with the boundary
of the shape.

Double-click
inside the original
shape to complete
the new shape.

7. Select the Autocomplete tool from the Draw tool drop-down list.

8. Click inside the existing cinder cone shape. Extend the line outside the shape and draw the boundary of the lava flow, clicking to change direction in the line. When you're finished, overshoot the boundary of the cinder cone shape again and double-click inside it. ArcView creates a new shape by combining the line you drew with the edge of the existing shape.

If you go outside the View window with the Autocomplete tool, your polygon won't be completed and you'll have to draw it again.

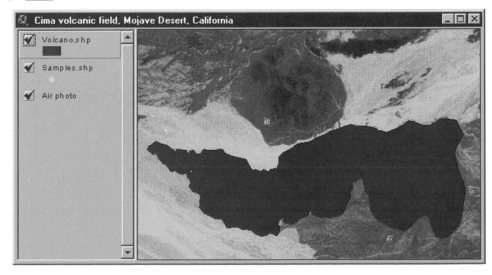

The Volcano.shp theme now contains two shapes that define the areas of the cinder cone and lava flow. The lava flow shape has handles around it because it's selected.

Snapping shapes. You can avoid leaving gaps between shapes by turning on ArcView's *snapping* option. When snapping is turned on, any new shape you draw will snap, or attach itself, to the nearest shape in the theme that lies within its *snapping tolerance.* To set snapping properties, click on the Editing icon in the Theme Properties dialog box. For more information, search for these Help Topics: *Setting a theme's snapping properties, Snap tools, What's new in creating and editing spatial data.*

9.   From the Theme menu, select Stop Editing. You're prompted to save your edits. Click Yes to save the shapes you've drawn to a shapefile.

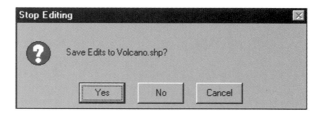

Notice that the dashed line around the Volcano.shp check box in the Table of Contents is gone. This means that the theme is no longer editable. (You can make it editable again by choosing Start Editing from the Theme menu.) The lava flow shape is highlighted because it was selected when you stopped editing.

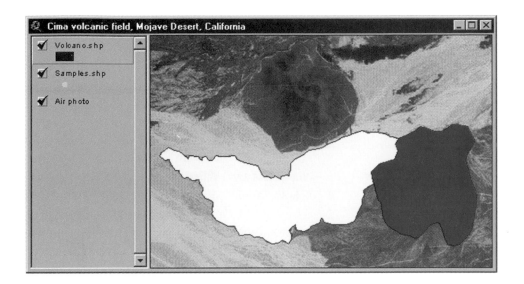

 10. Click the Clear Selected Features button to unselect the highlighted shape.

11. Drag the legend for the Samples.shp theme to the top of the Table of Contents. The Samples.shp theme redraws on top of Volcano.shp.

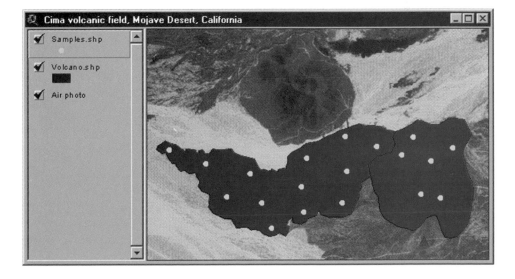

The outlines of the cinder cone and its lava flow are now stored as polygons in a shapefile.

Drawing point and line shapes

Suppose you want to add a few new sample points at places where you've recently collected rocks. You'd also like your map to show roads leading to the volcano.

Exercise 23c

1. If *ex23b.apr* is open, continue. Otherwise, choose Open Exercise from the File menu. In the Exercises scrolling list, select "ex23c," then click OK. When the project opens, you see a view with the air photo, sample points, and the polygons you drew in the last exercise.

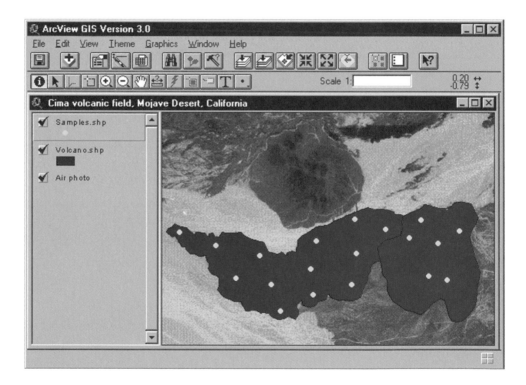

If you're opening exercise 23c, you may want to enlarge the ArcView and View windows.

ENVIRONMENTAL SYSTEMS RESEARCH INSTITUTE, INC.

2. Click on the Volcano.shp theme's check mark in the Table of Contents to turn off the theme.

3. Make sure the Samples.shp theme is active. From the Theme menu, select Start Editing. A dashed line appears around the theme's check box to show you that editing is enabled.

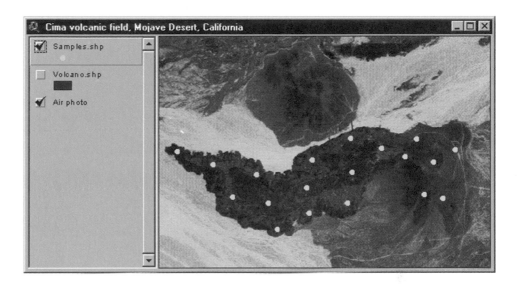

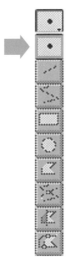

4. From the Draw tool drop-down list, select the Point tool.

5. Move the cursor over the cinder cone and lava flow. Click in a few different places to add three or four sample points. Every time you click, a new point is added. The most recently added point is selected.

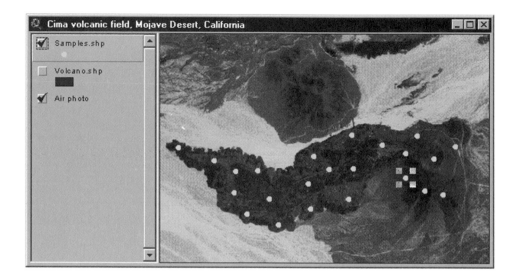

To undo your most recent edit, choose Undo Feature Edit from the Edit menu. By choosing this option repeatedly, you can undo (in reverse order) all edits made from the time you started editing.

6. From the Theme menu, select Stop Editing. Click Yes to save your edits.

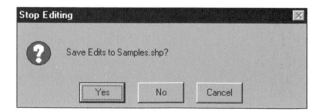

7. The selected point is now highlighted. Click the Clear Selected Features button to unselect it.

Now that you've added the sample points, you want to map the roads that lead from the desert floor to the cinder cone. In the lower right corner of the view is one that looks like a thin, curving scratch mark. You'll add a new line theme to the view and trace this road.

8. From the View menu, select New Theme. The New Theme dialog box opens. From the Feature type drop-down list, choose "Line," then click OK.

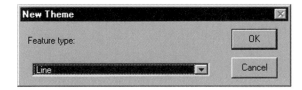

Another dialog box displays, asking you to name the theme.

9. Specify the *drive:\directory* where you want to save the new theme, then change the file name to **road.shp.** Click OK.

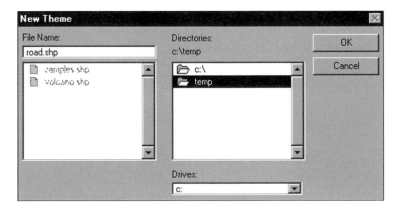

A new theme, Road.shp, is added to the view. The dashed line around its check box means it can be edited.

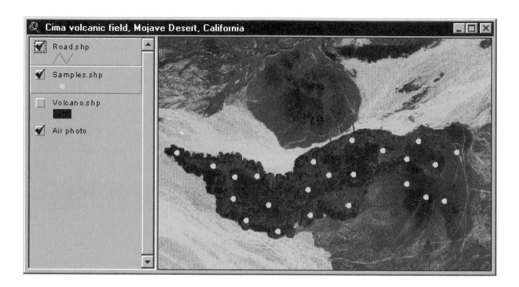

10. From the Draw tool drop-down list, select the Line tool.

11. Draw the line as shown on the next page. Click once to start the line and again to change direction. Double-click to end the line. (Your line doesn't have to look just like the one in the picture. If you don't like the line you draw, press the Delete key and start again.)

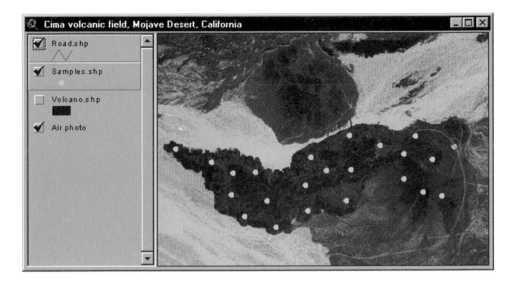

12. From the Theme menu, select Stop Editing. Click Yes at the prompt to save your edits.

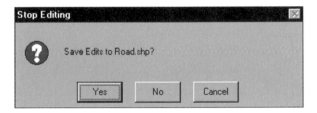

13. The line is highlighted. Click the Clear Selected Features button to unselect it.

If you want, you can bring up the Legend Editor and change the color or thickness of the line. (In the following graphic, the line thickness has been changed from one point to two.)

14. Click on the Volcano.shp theme's check box to turn it on.

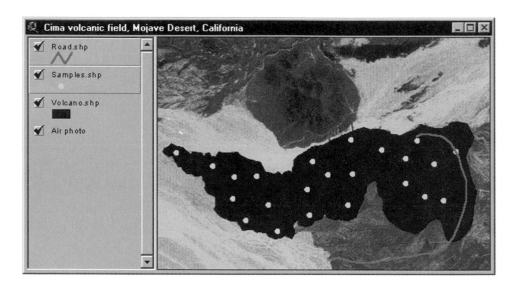

You've created three new shape themes using two different methods. You created the Samples.shp theme by converting features from an existing theme, then you edited it by adding point shapes. You created the Volcano.shp and Road.shp themes by drawing polygon and line shapes in new themes. As your study progresses, you can draw additional structures on this base map.

Before printing your map, you would put some finishing touches on it. You could give the Volcano.shp theme a transparent fill pattern to let the air photo show through. You could rename the themes and label the cinder cone and lava flow in the view.

You might also add new fields to the Samples.shp theme table to store the results of tests you run on the rock samples. These fields could then be used to classify the sample points.

If you want to go on to the next chapter, leave ArcView running. Otherwise, choose Exit from the File menu.

SECTION 7:
Creating your own data

Editing shapes
in a theme

Editing vertices

Merging and splitting shapes

Editing shapes in a theme

In chapter 23, you created shapes from existing features and drew new shapes using the Draw tool. Now you'll edit shapes by adding, deleting, and moving their vertices (a vertex is the point where two sides of a shape meet). You'll also learn how to split and merge shapes, and how to set rules for updating the shape attributes afterward.

Editing vertices

In ArcView GIS, you can edit a shape without affecting neighboring shapes, or you can edit a common boundary so that all shapes sharing that boundary are changed.

Suppose your company sells products door-to-door and you've divided the city into five sales territories by creating five polygon shapes on a map in ArcView. You need to update the sales territory map to include new streets on the outskirts of town. The territory to which the new streets are added will become disproportionately large, so you'll redraw the boundary it shares with another territory.

Exercise 24a

1. If necessary, start ArcView. From the File menu, choose Open Exercise. In the Exercises scrolling list, select "ex24a," then click OK. When the project opens, you see a theme showing the streets of the city and a theme showing the sales territories.

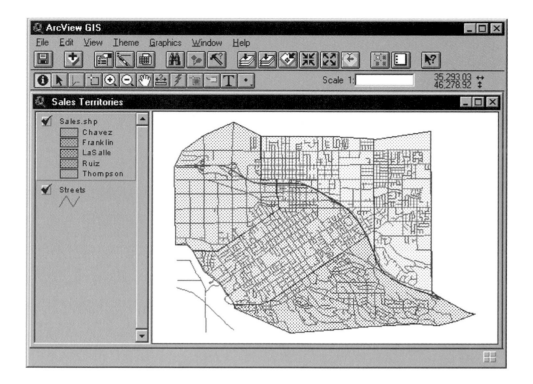

The Sales.shp theme is classified by the name of the salesperson assigned to each territory. Because the Sales.shp theme is created from a shapefile, it can be edited. The new streets are in the lower left corner of the view (the ones not covered by a sales territory shape). You decide that Gloria Ruiz will be the salesperson assigned to the new streets, as the streets are closest to her territory. You'll first edit Ruiz's territory without affecting the adjacent territories.

2.   Make sure the Sales.shp theme is active by clicking on it in the Table of Contents. From the Theme menu, select Start Editing. A dashed box appears around the Sales.shp theme check box to show that it's editable.

3.   Use the Zoom In tool to draw a box around the lower left corner of the view ArcView will zoom in to this area when you release the mouse button.

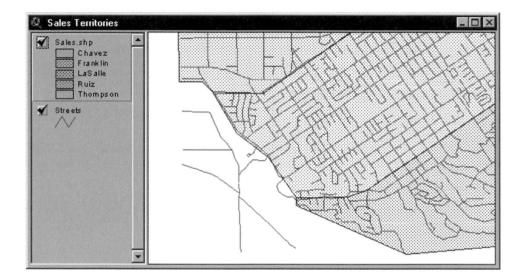

Now, you'll change the boundary of the sales territory by moving some of its vertices and adding new ones.

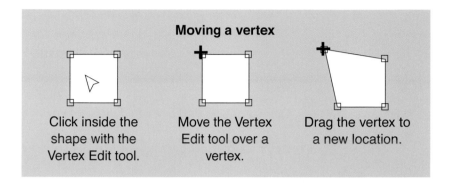

Moving a vertex

Click inside the shape with the Vertex Edit tool.

Move the Vertex Edit tool over a vertex.

Drag the vertex to a new location.

4. Click on the Vertex Edit tool, then select Ruiz's sales territory by clicking inside the pink shape.

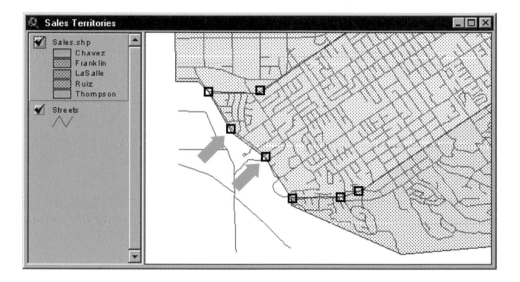

A square selection handle appears over each vertex of the shape, indicating that the vertex can be moved or deleted. (The handles have been slightly enlarged in this book so you can see them better.) The green arrows point to the two vertices you're going to move in the next steps. (You won't see the green arrows on your screen.)

5. Place the cursor over one of the vertices indicated by a green arrow in the previous step. The cursor changes to crosshairs. Hold down the mouse button and drag the vertex to its new location (shown by the green arrows below). Drag the second vertex to its new location as well. (If you mistakenly drag the wrong vertex, you can select Undo Edit from the Edit menu to undo the change you just made.)

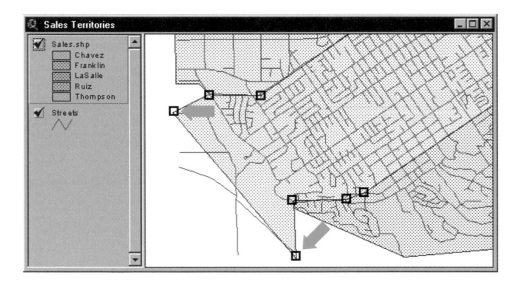

To include the remaining streets, you'll add a new vertex to the polygon and drag it to a new location.

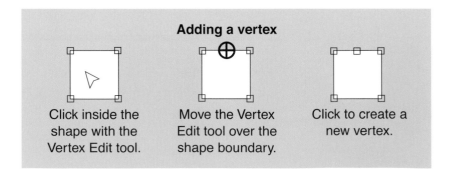

Adding a vertex

Click inside the shape with the Vertex Edit tool.

Move the Vertex Edit tool over the shape boundary.

Click to create a new vertex.

6.   Move the Vertex Edit tool over the boundary at the location shown below by the green arrow. The cursor changes to a target (a plus sign inside a circle). Click to add a new vertex at that location.

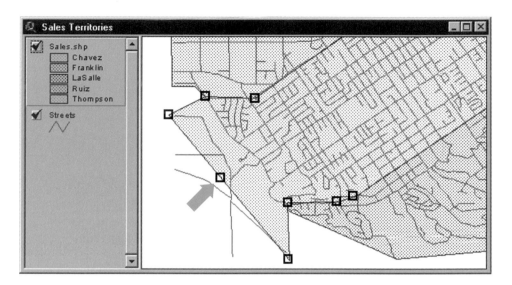

7.   Move the Vertex Edit tool over the new vertex until the cursor changes to crosshairs. Drag this vertex toward the left corner of the screen until all of the new streets are covered by the shape.

The new streets are now completely included in Ruiz's sales territory. You decide to straighten the territory's boundary by deleting a vertex.

8. Move the Vertex Edit tool over the vertex shown by the green arrow and press the Delete key.

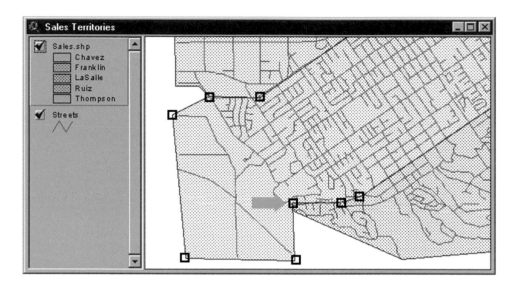

The vertex disappears and the shape looks as shown below.

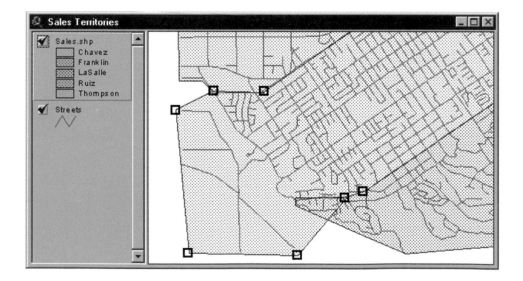

This isn't the effect you wanted. Instead of shifting the boundary between Ruiz's and LaSalle's territories, your edit has made the two territories overlap. You decide to undo the edit.

9. Click and hold the right mouse button. A popup menu appears with editing choices. Choose Undo Feature Edit. The deleted vertex reappears.

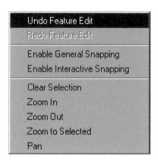

The Vertex Edit tool can be used to edit one shape independently of its neighbors or to edit the common boundaries and vertices between shapes. When you click *inside* a shape, adding, deleting, or moving a vertex changes only the selected shape (this may result in gaps or overlaps between a shape and its neighbor). When you click the Vertex Edit tool *directly on a common boundary* between two shapes, you can only edit vertices that lie on that boundary. Adding, deleting, or moving a vertex adjusts both shapes. When three or more shapes meet at a vertex, you can click the Vertex Edit tool *directly on the common vertex* to select only that vertex. When you move the vertex, all the shapes are changed.

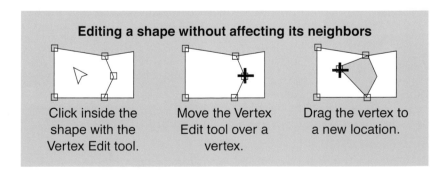

Editing a shape without affecting its neighbors

| Click inside the shape with the Vertex Edit tool. | Move the Vertex Edit tool over a vertex. | Drag the vertex to a new location. |

Because Ruiz's territory has become too large, you'll give some of her territory to LaSalle by editing the common boundary between the two territories.

10. Click and hold the right mouse button. From the popup menu, choose Zoom to Selected to see all of Ruiz's territory. Zoom to Selected zooms to the extent of the selected shape.

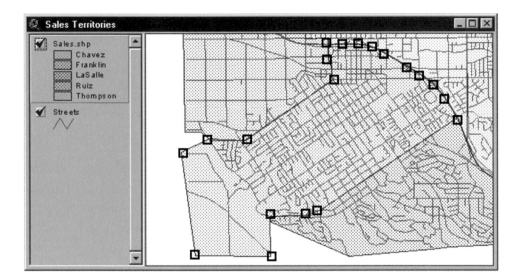

You can see that Ruiz's territory is still selected—every vertex has a square selection handle on it. You want to edit the common boundary between the Ruiz and LaSalle territories, so you'll deselect the Ruiz territory and select the common boundary.

11. Click on the white area inside the view but outside the shapes. The vertex handles disappear because the shape is no longer selected.

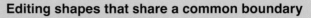
Editing shapes that share a common boundary

To change both shapes that share a common boundary, click the Vertex Edit tool directly on the boundary. The vertices that can be moved have a square handle. Vertices that can't be moved have a circular handle and are called *anchor points*.

Click directly on the common boundary.

Move the cursor over the vertex.

Drag the vertex to a new location.

12. Move the Vertex Edit tool over the common boundary line between the Ruiz and LaSalle territories and click. You see circular, non-movable anchor points at the ends of the common boundary line and square handles on the vertices in between. (There should be two round anchor points and two square handles, marked by the green arrows in the following graphic. If you see more selection handles than these, you've selected one of the shapes instead of the common boundary. Click on a white area to unselect the shape and click again on the common boundary.)

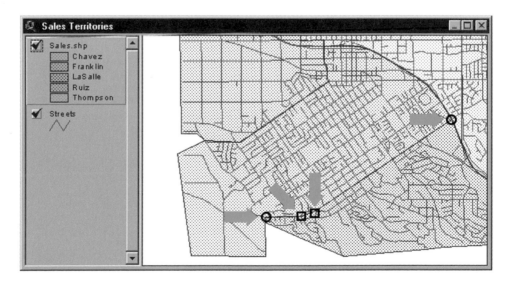

13. Click and drag the two square handles (one at a time) to the positions shown in the following graphic. The exact placement doesn't matter.

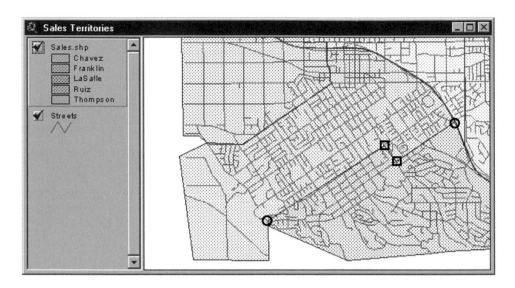

Editing shapes that meet at a vertex

To edit three or more shapes that share a common vertex, when you want all the shapes to change, click directly on the vertex with the Vertex Edit tool. A square handle appears on the movable vertex and circles appear on the nonmovable anchor points.

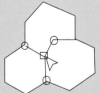

Click directly on the common vertex.

Drag the vertex to a new location.

14. Click the Zoom to Full Extent button to view the entire sales territory map. Click on any white space to clear the selection handles.

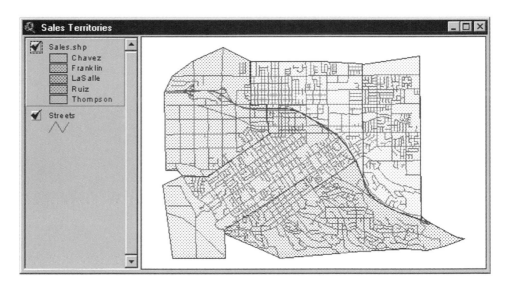

You have edited a single shape and a common boundary between shapes to define the territories the way you want. The new sales territory map includes all the streets with no overlapping territories.

Digitizing in ArcView. The version of ArcView that comes with this book teaches you how to edit shapes with a mouse, but the standard version of ArcView also supports a digitizer (a tablet with a pointing device) that lets you edit shapes more precisely. To use a digitizer with ArcView, you need to load the ArcView Digitizer extension. (For details on extensions, see chapter 7.) You also need a supported driver for the digitizer. For details on setting up a digitizer and a list of supported digitizers, search for these Help Topics: *Digitizer extension (description), Digitizer setup (Dialog box), Supported digitizers.*

15. From the Theme menu, select Stop Editing. When you're prompted to save your changes, click Yes.

Merging and splitting shapes

Suppose that one of your salespeople, Larry Thompson, quits. You decide to divide his territory between two other salespeople until you can hire someone new. After splitting Thompson's territory, you'll merge each resulting part with one of the other territories, leaving you with four sales territories.

For each sales territory, there's a Sales_Target attribute in the theme table containing the yearly sales goal for that territory. What will happen to this attribute when you split Thompson's territory and merge it with the others? When you split and merge shapes, you need to tell ArcView how to update the attributes of the affected records. If you split a territory in two, ArcView creates a new record in the attribute table because there is now an additional shape. If you merge two territories, ArcView deletes a record from the table because there is one less shape. You'll set up rules for reapportioning the Sales_Target attribute as the number of records in the table changes.

Exercise 24b

1. From the File menu, choose Open Exercise. In the Exercises scrolling list, select "ex24b," then click OK. When the project opens, you see a view with the Streets theme underneath the Sales2.shp theme showing the sales territories. You also see the attribute table of the Sales2.shp theme.

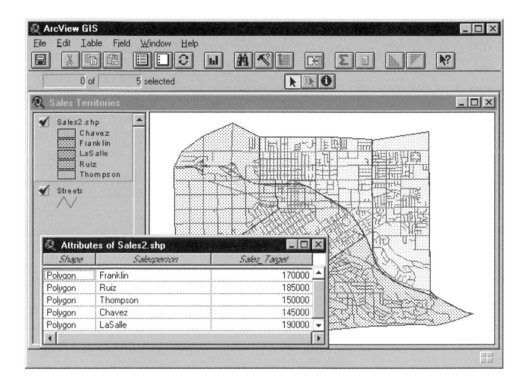

You see five records, one for each sales territory. Notice that the amount in the Sales_Target field for Thompson, the salesperson who is leaving, is $150,000. You want this number to be apportioned by area when you divide his sales territory into two parts.

2. Make the view active by clicking on its title bar. Make sure the Sales2.shp theme is active by clicking on it in the Table of Contents. From the Theme menu, choose Start Editing. A dashed line appears around the check box to indicate that the Sales2.shp theme is editable.

3. From the Theme menu, choose Properties. The Theme Properties dialog box displays.

4. Click on the Editing icon on the left side of the Theme Properties dialog box to display editing properties.

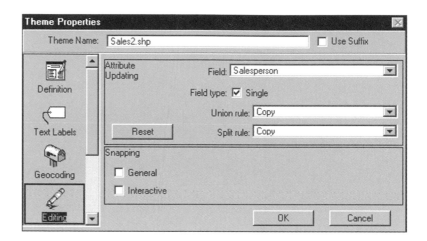

In the Field drop-down box, you see that the Salesperson field is selected. Below it are the Union rule and Split rule drop-down boxes. You use these boxes to determine what will happen to the Salesperson attribute when you split or merge the shape. By default, both the Union rule (for merging shapes) and the Split rule are set to Copy. When you select two shapes and merge them, the records for the original shapes are deleted from the attribute table and a new record is created for the resulting shape. The Salesperson attribute for the new record will be copied from the original record that occurs first in the attribute table. If you split a shape, the record for the original shape is deleted and two records are created for the new shapes. The Salesperson attribute from the original record is copied into the Salesperson attributes for each of the new records in the attribute table.

You'll leave the Union and Split rules set to Copy for the Salesperson attribute, but you want to change them for the Sales_Target attribute.

5.   From the Field drop-down list, choose "Sales_Target."

When you merge these new territories with others, you want the Sales_Target values of the merged territories to be added together.

6.   From the Union rule drop-down list, choose "Add."

When you split the territory, you want the Sales_Target value to be divided proportionately by area for the two new territories.

7.   From the Split rule drop-down list, choose "Proportion."

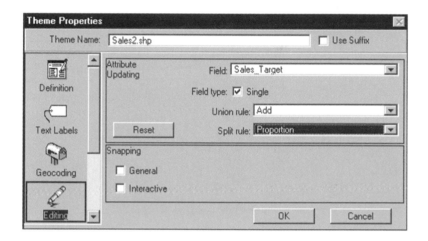

8.   Click OK.

Now that you've set the Union and Split rules, you'll edit the theme.

Using the Split tool

The Split tool divides one shape into two. When you use the Split tool to draw a line that completely bisects a shape, the line becomes the boundary between two new shapes.

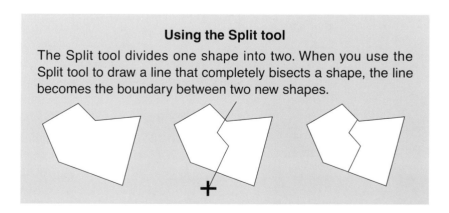

9. Select the Polygon Split tool from the Draw tool drop-down tool bar.

10. Click outside the shape you want to split (Thompson's, the blue one) to begin drawing a split line. It's important that your split line completely split the shape, so you'll extend it beyond the shape boundaries. Follow the vertical street down the middle of the shape, as shown by the arrow in the following graphic. Extend the line beyond the shape boundary and double-click to end the line. (Your line doesn't have to look exactly as shown.)

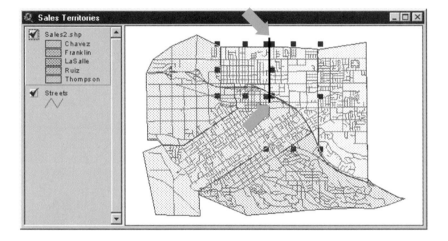

The shape is split in two and both resulting shapes are selected.

Splitting line shapes. In addition to the Polygon Split tool, there's a Line Split tool for line themes. When you cross an existing line with a line drawn with the Line Split tool, both the existing line and the new line are split at the intersection. For more information, search for these Help Topics: *Splitting_line_features, Drawing_and_editing tools, What's new in creating and editing spatial data.*

11. Make the attribute table active again by clicking the Open Theme Table button.

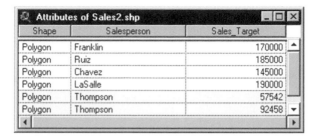

Notice that there are now six records in the table. There are two records for Thompson's territory, showing Sales_Target amounts of 57,542 and 92,458. (Your amounts may vary, depending on how you split the territory.) These add up to 150,000, the original value before you split the territory. The original amount was divided in proportion to the areas of the two new shapes. The Salesperson attribute was copied into both records because the Split rule for this attribute is still set to the default of Copy.

Finally, you want to add part of the split territory to Franklin's territory (on the left) and the other part to Chavez's territory (on the right).

 12. Make the view active again and click on the Pointer tool. Click anywhere in Franklin's territory (the red shape) to select it. Now hold down the Shift key and click on the left side of the subdivided Thompson territory. Both shapes are selected.

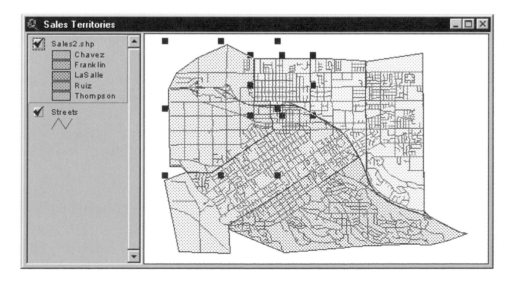

13. From the Edit menu, select Union Features. The two shapes are merged into a single sales territory.

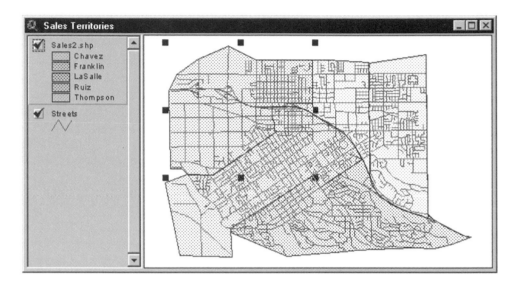

14. Click on Chavez's territory (the green shape) with the Pointer tool to select it. Hold down the Shift key and click on the remaining portion of Thompson's territory to select it as well.

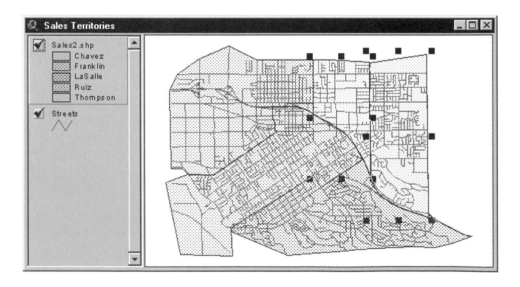

15. From the Edit menu, select Union Features. The two selected shapes are merged into one sales territory.

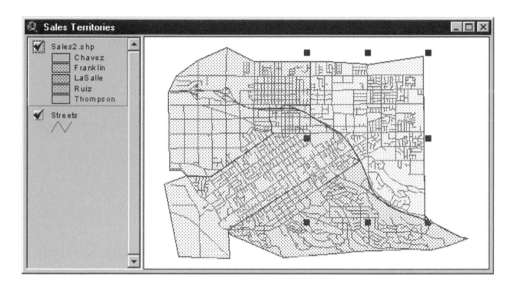

 16. Make the attribute table active again by clicking the Open Theme Table button.

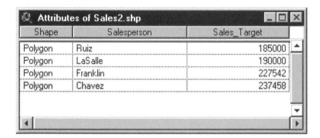

Notice that there are only four records. The Sales_Target for Franklin's territory has increased to 227,542. That's the sum of his original Sales_Target of 170,000 plus 57,542 from the merged part of Thompson's territory (again, your value may be slightly different). The Sales_Target value for Chavez's territory has also increased since it was merged. The Union rule for the Salesperson field is set to Copy. ArcView copied the name "Franklin" to the merged record because the record for Franklin occurred in the table before Thompson.

17. Make the view active and click on white space to unselect the selected shape.

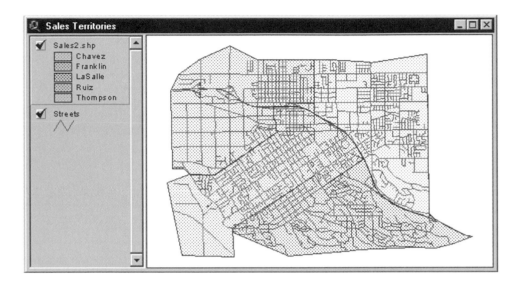

Your sales territory map now has four territories instead of five and the sales amounts for the territories have been adjusted according to your specifications.

18. Select Stop Editing from the Theme menu. When you are prompted to save your changes, click Yes.

If you want to go on to the next chapter, leave ArcView running. Otherwise, choose Exit from the file menu.

    ENVIRONMENTAL SYSTEMS RESEARCH INSTITUTE, INC.

CHAPTER 25

SECTION 7:
Creating your own data

Creating themes from coordinate files

Adding event themes

Creating themes from coordinate files

ArcView GIS lets you create themes from location information in a table, information in the form of geographic coordinates, street addresses, or mileposts along a route. Chapter 26 focuses on creating themes from addresses. This chapter focuses on creating themes from geographic coordinates.

Geographic coordinates can be expressed either as degrees of latitude and longitude, or as pairs of x,y coordinates (see chapter 5). ArcView reads the geographic coordinates from fields you specify in a table and creates a point feature for each location.

Geographic coordinates can be obtained from paper maps, locations in a view, field surveys, Global Positioning System (GPS) receivers, and so on. The geographic coordinates can be in any tabular data format that ArcView supports.

Adding event themes

In ArcView, locations stored in a tabular format are referred to as *event locations* or simply *events,* and the table containing them is referred to as an *event table*. Events let you map data that contains geographic locations but isn't in a spatial format (e.g., a file of addresses, a table of information referenced to milepost locations along a route, or latitude–longitude locations stored as records in a table).

Suppose you're involved in a conservation program designed to help endangered African wildlife and develop local economies. As a part of this program, permits are sold to foreign hunters allowing them to hunt certain species of African wildlife, in limited numbers. The money so collected goes to local villages for building schools and hospitals. In return, villagers agree to monitor protected areas and prevent poaching inside these areas.

During the last month, more than twenty antelope have fallen prey to poachers inside protected areas. You want to know exactly where these incidents took place and which protected areas they occurred in.

You've sent inspectors out, armed with portable GPS equipment, to capture precise x,y locations. The x,y locations are measured in decimal degrees (degrees of latitude and longitude expressed as a decimal), where x is the longitude and y is the latitude. You've received the data in a dBASE-formatted file.

Exercise 25a

1. If necessary, start ArcView. From the File menu, choose Open Exercise. In the Exercises scrolling list, select "ex25a," then click OK. When the project opens, you see a view with two themes, Villages and Protected Areas. Each village is a point; each protected area is a polygon defining a conservation unit. There's one conservation unit for each village. Beyond the Village Protection areas is a large Federal Protection area.

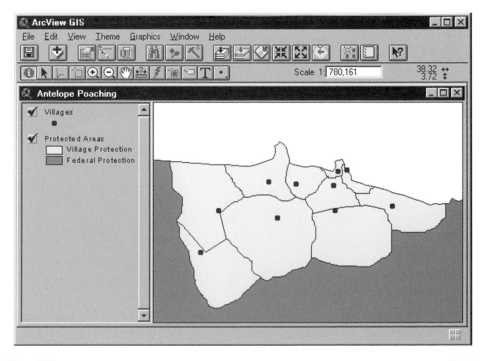

You'll bring the dBASE file containing x,y locations into the current project as an ArcView table.

2. From the Window menu, make the Project window active, then click on the Tables icon.

3. Click the Add button to open the Add Table dialog box. From the lower left drop-down list (List Files of Type), choose "dBASE [*.dbf]." This indicates you want to create a table from a dBASE file. From the Drives list, select the drive where you installed the data for this book, then navigate to \gtkav\data\ch25 in the Directories list. From the File Name list, choose "antelope.dbf."

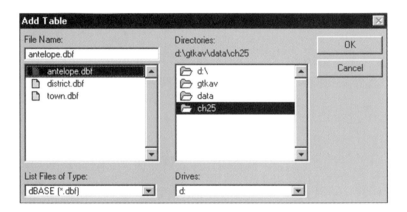

4. Click OK to add the antelope.dbf table to your project.

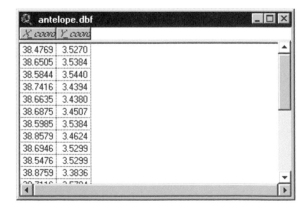

When the table opens, you see two fields, *X_coord* and *Y_coord*. You'll use the location coordinates in these fields to create a new theme of point features based on ArcView's shapefile format. Since the table doesn't need to remain open, you'll close it first.

5. Close the antelope.dbf table. Make the view active, then choose Add Event Theme from the View menu. The Add Event Theme dialog box displays.

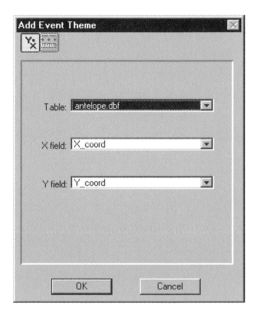

At the top of the Add Event Theme dialog box, you see two buttons. Each button represents a category of events: XY (selected) or Route. (Clicking a button displays the fields appropriate for the category you choose.)

Antelope.dbf is already selected in the Table list. ArcView reads the field names in this table to find fields likely to contain x,y coordinates. The names of these fields, *X_coord* and *Y_coord,* appear in the X field and Y field lists.

6. Click OK to create a new theme from the x,y coordinates in the antelope.dbf table.

The new theme, Antelope.dbf, appears in the view's Table of Contents.

7. Click on the check box in front of the theme name to turn it on. ArcView draws the view with the new theme.

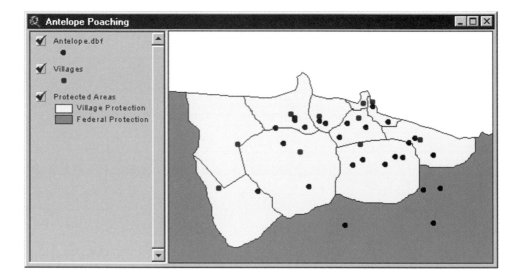

Now you can see exactly where the antelope poachings took place.

For each poaching site, you want to know the name of the protected area it's in. So you'll perform a spatial join.

8. Make the Antelope.dbf and Protected Areas themes active by holding down the Shift key and clicking on each theme, then click the Open Theme Table button to open the attribute tables for both themes.

9. Make the Attributes of Protected Areas table active, then click on its Shape field to make it active. (This is the source table.) Do the same to the Attributes of Antelope.dbf table. (This is the destination table.)

ENVIRONMENTAL SYSTEMS RESEARCH INSTITUTE, INC.

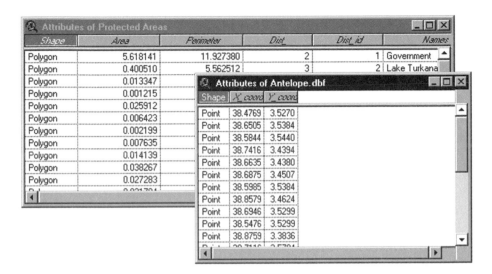

10. Make sure the Attributes of Antelope.dbf table is active, then select Join from the Table menu. ArcView appends the attributes of each protected area to the poaching sites it contains. (See chapter 20 for more information on spatial joins.)

The Attributes of Antelope.dbf table displays with additional fields appended from the Attributes of Protected Areas table.

11. Widen the table so you can see all the fields.

For each antelope poaching site, there's now a name, the name of the protected area it's found in.

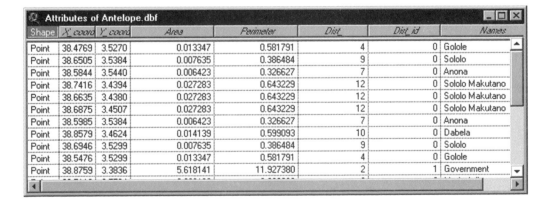

Now that you know the names of the protected areas where poaching incidents occurred, you could use Summarize (chapters 14 and 15) to find out how many antelope were lost in each protected area and which villages need to be more watchful.

If you want to go on to the next chapter, leave ArcView running. Otherwise, choose Exit from the File menu.

SECTION 7:
Creating your own data

Address geocoding

Making a theme matchable

Matching a list of addresses

Handling unmatched addresses

CHAPTER 26

Address geocoding

In the last chapter, you created points on a map from a table of x,y coordinates. In this chapter, you'll use *address geocoding* to create points on a map from a table of addresses.

To geocode addresses, you need a *reference theme* to serve as a base map (the reference theme is usually a street theme with attributes that specify the street name, street type, and range of addresses that occur along each street). You also need an *address table* that contains the addresses you want to locate as points on the map.

The geocoding process consists of three steps. First, you prepare the reference theme for geocoding (this is called making the reference theme *matchable*). Next, you *batch match* addresses. During this step, ArcView GIS compares the table of addresses to the street information in the reference theme and looks for matches. Each matched address is added as a point in a new *geocoded theme*. In the third step, you *re-match* the addresses that couldn't be batch matched, and ArcView adds these points to the geocoded theme.

In this chapter, you'll make a street theme matchable, then match a table of addresses to create a geocoded theme. You'll also re-match addresses that fail to match. Finally, you'll see how to match an address that's not in an address table.

Making a theme matchable

The first step in geocoding is to make the reference theme matchable by selecting an *address style*. An address style determines which address components (e.g., street name, street type, direction, ZIP Code) will be required elements in the address-matching process. For instance, you could choose an address style in which addresses can be matched by street name and number alone, or you could require that additional components, such as street direction or ZIP Code, be present. The address style you choose will depend mainly on the fields that are available in the reference theme table.

Suppose that you work for Ellsworth's Office Supply Company. It's your job to assign sales territories based on the distribution of thousands of customers across the state. Traditionally, you print out a customer list from a database, take a box of pushpins to the wall map, and attempt to locate each address manually. With ArcView, you'll be able to accomplish in minutes what used to take weeks. Because you've never geocoded before, you decide to make a test run with a sample of your customer list. First, you'll make a small street theme matchable so ArcView can use it for geocoding.

Exercise 26a

1. If necessary, start ArcView. From the File menu, choose Open Exercise. In the Exercises scrolling list, select "ex26a," then click OK. When the project opens, you see a view with a street theme for a portion of the Atlanta area. You also see the theme attribute table and the customer.dbf address table. (The customer.dbf table contains the list of customer addresses you want to geocode.)

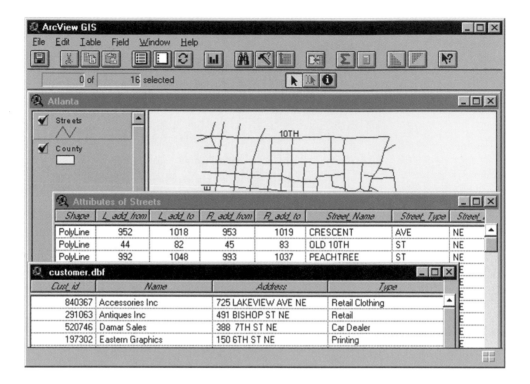

First, look at the Attributes of Streets table. The addresses in this table are divided into components. The range of address numbers on each street is stored as four fields in the table. The fields *L_add_from* and *L_add_to* contain the lowest and highest address numbers on the left-hand side of the street; *R_add_from* and *R_add_to* contain the lowest and highest on the right-hand side. There's also a *Street_Name* field, a *Street_Type* field, and a *Street_Dir* field. You'll select an address style that uses these fields for matching.

Now look at the customer.dbf table, which contains the addresses you want to locate. This table is not a theme attribute table because it's not associated with any theme: it was added to the project from the Table icon in the Project window. Each address contains a single address number. Notice that the addresses in this table are *not* divided into components. ArcView will divide them into components later.

2.   Make the view active. From the Theme menu, choose Properties to display the Theme Properties dialog box.

3.   Click on the Geocoding icon on the left to display the geocoding theme properties.

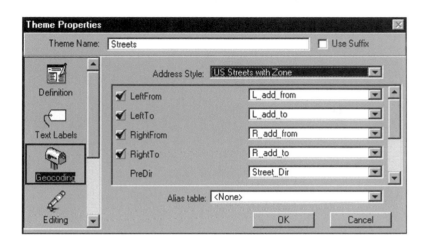

Notice that ArcView selects "US Streets with Zone" in the Address Style field. It chooses this address style because there are fields in the reference theme table that correspond to each of the required address components for this style. Under the Address Style field, there's a large scrolling box. On the left side is a list of address components, such as *LeftFrom* and *LeftTo*. Components required by the US Streets with Zone address style have check marks next to them. (If there is no check mark, the component is optional.) On the right side of the box is the name of the field in the reference theme table that corresponds to the address component. ArcView selects the field name most likely to contain the correct information. (If necessary, you can use the drop-down arrow to change the selection.) For example, ArcView has determined that the *L_add_from* field in the reference theme is most likely to contain the *LeftFrom* address component.

4.   Scroll down to see all the address components and their corresponding fields. Notice that the LeftZone and RightZone components are

required for this address style. The fields corresponding to these components are the ZIP Code fields in the reference theme.

Because the test addresses in the customer.dbf table (the table of addresses you're going to match to the reference theme) don't have ZIP Codes, you want an address style that doesn't require ZIP Codes. The fewer address components ArcView must match, the faster it geocodes, although the geocoded points may be located less precisely.

5. From the Address Style drop-down list, select "US Streets." Scroll through the list of components so you can see the requirements for this address style.

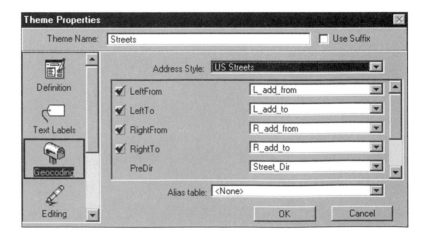

The US Streets address style requires address components such as *LeftFrom* and *RightFrom* (the lowest address numbers for the street), and *LeftTo* and *RightTo* (the highest address numbers). There are no ZIP Code components in this style.

The field names for the required address components are set correctly, so you don't need to change them. You'll set the optional PreDir (prefix direction) component to "None," because the customer.dbf table you're going to match with the reference theme doesn't have prefix directions before the street names.

6. From the PreDir drop-down list, select "<None>." ArcView will ignore this address component during address matching.

7. Click OK to set the geocoding theme properties. The Build Geocoding Index dialog box displays.

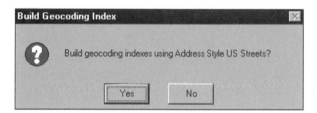

8. Click Yes. ArcView builds geocoding indexes for the address fields in the attribute table to make address geocoding faster.

The Streets reference theme is now matchable. You only have to make the reference theme matchable once unless you want to change the address style or the fields used for each address component.

Matching a list of addresses

The second step in address geocoding is to batch match addresses in a table to the reference theme. ArcView completes two processes during batch matching. The first is *address matching,* where addresses in the address table are compared to address ranges in the reference theme. The second is *geocoding,* where each matched address in the address table is assigned geographic coordinates and made a point feature in a new theme.

Now that you have a matchable theme, you'll geocode the customer.dbf address table, which represents a small portion of Ellsworth's customer database.

Exercise 26b

1. If *ex26a* is open, continue. Otherwise, choose Open Exercise from the File menu. In the Exercises scrolling list, select "ex26b," then click OK. When the project opens, you see a view with the Streets theme.

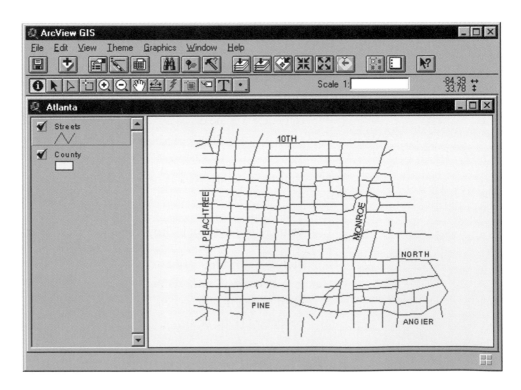

To set up the address-matching process, you use the Geocode Addresses dialog box.

2.   Choose Geocode Addresses from the View menu. The Geocode Addresses dialog box displays.

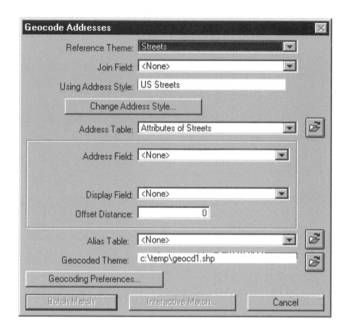

The Reference Theme drop-down list shows the reference theme, Streets. The Using Address Style box shows the address style you selected for the reference theme. The Address Table list lets you select the address table you want to geocode. You'll need to change this to the customer.dbf table.

3.   From the Address Table drop-down list, choose "customer.dbf."

To match every address in the table automatically, you'll use Batch Match. A geocoded point theme of matched addresses will be created and added to the view. The theme will be saved to the location shown in the Geocoded Theme box.

4.   Click Batch Match to start address matching.

ArcView matches addresses the same way you do. When you want to find 532 N. Central Avenue on a map, you break the address down into parts and search for each part. First, you find a street on the map called Central. You check the street type to make sure it's Central Avenue, not Central Boulevard, and you check the street direction to make sure you're looking at the north section of the street, not the south. Finally, you scan up and down the street numbers on the north section of the street until you find the block where 532 is located.

To make these same comparisons, ArcView divides each address in the address table into components (this process is called *standardization*). Then it tries to match components in the address table with components in the reference theme table. An address record in the reference theme table becomes a *match candidate* if it resembles a given address in the address table. ArcView assigns the candidate a match score, based on the degree of resemblance. ArcView then looks at the candidate with the highest match score. If its score is above the *minimum match score*, ArcView matches it and geocodes a point location. You'll see these steps in more detail when you re-match addresses manually.

ArcView tries to match each address in the customer.dbf table. When it's done, the Re-match Addresses dialog box displays.

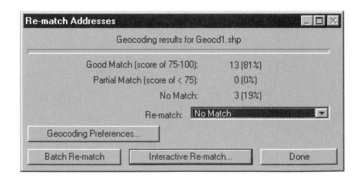

The Re-match Addresses dialog box informs you that thirteen of the six-teen records in the table had good matches, with scores of 75 or above, but that three records weren't matched.

Understanding match scores. Each component of an address record in the reference theme table is scored based on how closely it matches the corresponding component in the address table. The scores of the components are then weighted by importance (for example, street name is more important than street direction) and added together to get a total score. A perfect score of 100 occurs only if each part of the address matches exactly. If ArcView compares the street name "Coton-wood" in the address table to "Cottonwood" in the reference theme, it will assign a less-than-perfect score to the street name component in the reference theme table. If ArcView compares "Redwood" and "Ave" to "Redwood" and "Blvd," it will assign a perfect score to the street name, but a score of 0 to the street type. ArcView considers a total weighted score of 75 or above to be a good match; a score below 75 but above the minimum match score counts as a partial match. (The default minimum match score is 60, but you can change it.) For more information, search for these Help Topics: *Address matching, Geocoding Editor (Dialog box), Overview of address geocoding.*

You'll re-match the unmatched addresses later; for now, you'll conclude address matching and look at the geocoded theme ArcView creates.

5.   In the Re-match Addresses dialog box, click Done. ArcView creates a shapefile that contains a point feature for each of the thirteen matched addresses. The shapefile is added to the view as a theme, Geocd1.shp.

6. Click on the Geocd1.shp theme to make it active, and display it by clicking on its check box. (The color symbol for your theme will probably be different, as it's randomly selected.)

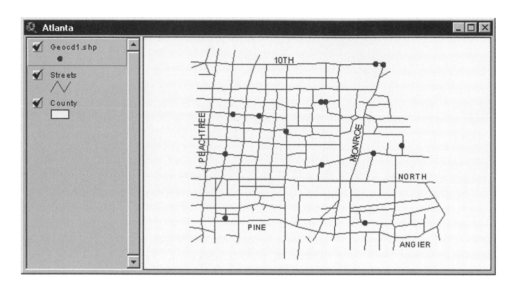

Geocoding the address. The US Streets address style has left and right street number ranges as required address components. Interpolation is used to locate an address along the length of the street and place a point at the correct location. For instance, 230 SUNSET ST (a record from the address table) matches to a record with the address range 201 299 200 298 SUNSET ST (from the reference theme table). Because the address 230 falls about one-third of the way between the lowest address (201) and the highest (298), ArcView places a point about one-third of the way along the street. Because the left street addresses are odd numbers and the right street addresses are even, ArcView places the point on the right side of the street. You can exaggerate the offset from the street by setting an offset value in the Geocoding Addresses dialog box. For more information, search for these Help Topics: *Geocoding Editor (Dialog box), Overview of address geocoding.*

7. With the Geocd1.shp theme active, click the Open Theme Table button to open the theme attribute table. Widen the table and scroll to the right to see all the fields in the table.

Address	Type	Av_add	Av_status	Av_score	Av_side
500 RANKIN ST NE	Real Estate	500 RANKIN ST NE	M	100	L
751 JUNIPER	Wholesale	751 JUNIPER	U	0	
555 10TH ST NE	Stockbroker	555 10TH ST NE	M	100	R
220 6TH ST NE	Trade School	220 6TH ST NE	M	100	L
400 7TH ST NE	Retail Clothing	400 7TH ST NE	M	100	L
844 MIRTEL ST NE	Retail Food	844 MIRTEL ST NE	U	0	
711 KENNESAW AVE NE	Wholesale	711 KENNESAW AVE NE	M	100	R
991 MONROE DR NE	Restaurant	991 MONROE DR NE	M	100	R
400 PONCE DE LEON AVE NE	Retail Entertainment	400 PONCE DE LEON AVE NE	M	100	L
699 JUNIPER ST NE	Real Estate	699 JUNIPER ST NE	M	100	R
42 NORTH NE	Bookkeeping	42 NORTH NE	U	0	

ArcView copied all the fields from the original customer.dbf table into the attribute table for the Geocd1.shp theme. Some new fields were also created as a result of the address-matching process. The Av_add field shows the address used during matching. (This field contains the same information as the Address field, but can be edited during the re-match process. See the boldfaced note on page 26-18.) Av_status indicates matched addresses with an "M" and unmatched addresses with a "U." Av_score shows the match score. Av_side indicates whether the address is on the right or left side of the street. Your original customer.dbf address table is unchanged.

8. Close the theme table.

If you want to continue, leave the exercise open.

Handling unmatched addresses

The last step in address geocoding is to re-match, interactively, the addresses ArcView couldn't batch match. Addresses may fail to match because of misspellings, nonexistent street numbers, missing address parts, or other errors. (Errors may exist either in the address table data or in the reference theme table data.) Interactive re-matching lets you look at the candidate addresses from the reference theme and decide if any are close enough to accept as matches. You can also fix spelling errors, adjust ArcView's spelling sensitivity, or lower the minimum match score to increase the likelihood of finding a candidate to match.

Exercise 26c

1. If *ex26a.apr* or *ex26b.apr* is open, continue. Otherwise, choose Open Exercise from the File menu. In the Exercises scrolling list, select "ex26c," then click OK. You see a view of the Streets theme and the geocoded theme you just created, Geocd1.shp. The theme displays the thirteen matched address locations from the previous exercise.

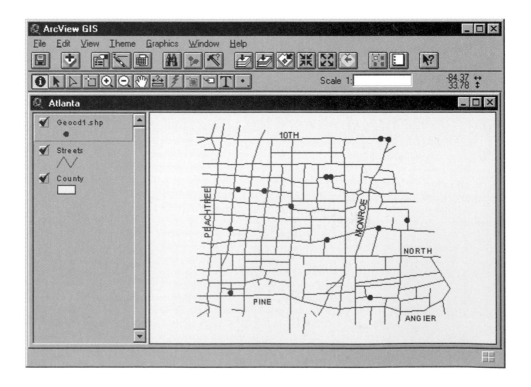

    ENVIRONMENTAL SYSTEMS RESEARCH INSTITUTE, INC.

2. From the Theme menu, choose Re-match Addresses. The Re-match Addresses dialog box displays, showing that there are three unmatched records.

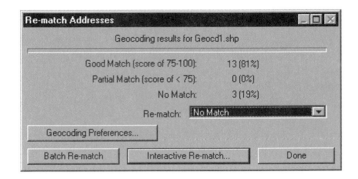

By default, the Re-match drop-down list is set to No Match, which means you'll re-match only those addresses that couldn't be batch matched. You could also choose to re-match any partial matches (if you wanted to find out why they didn't match exactly), or you could re-match all records.

3. Click the Interactive Re-match button. The Geocoding Editor dialog box displays.

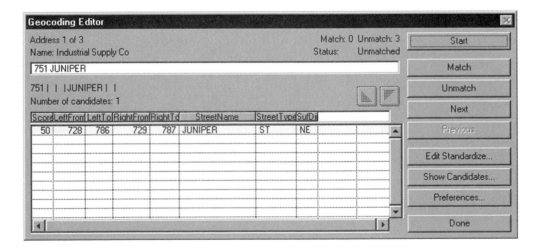

You see the first unmatched address from the address table, "751 JUNIPER," in the text box at the top of the Geocoding Editor. Beneath this is the address in its standardized form, with vertical bars showing how ArcView divided it into the components "751" and "JUNIPER." The scrolling list shows one candidate from the reference theme table. During the batch match operation, it received a match score of 50. (The score appears in the first field of the scrolling list.) By default, any address record in the reference theme with a match score of 30 or better qualifies as a candidate. An address record with a score of 60 or better is a partial match, and an address record that scores 75 or above is a good match. Good and partial matches are geocoded. This candidate falls below the minimum score for a partial match, so it isn't geocoded.

Although the address at the top of the Geocoding Editor dialog box lacks a street type and a suffix direction, you decide that the candidate should be matched to it.

4.   Click the Match button. Even though the candidate score is below the minimum match score of 60, ArcView forces a partial match between the address and the candidate and displays the next unmatched address.

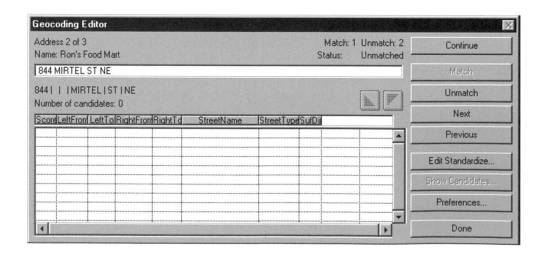

Note that there are no candidates for the current unmatched address (none of the address records in the reference theme meets the default minimum candidate score of 30).

ArcView uses a set of criteria, called *geocoding preferences,* to determine whether an address record is a candidate for a match. Because no candidates were found for 844 MIRTEL ST NE, you'll relax some of the geocoding preferences.

5.   Click the Preferences button in the Geocoding Editor. The Geocoding Preferences dialog box displays.

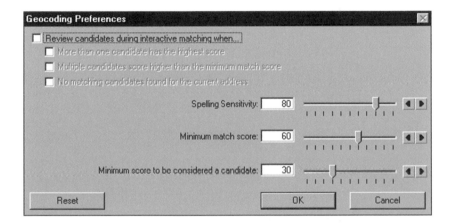

You suspect that "Mirtel" is a misspelled street name, so you decide to lower the spelling sensitivity. This will allow address records with a greater variation in spelling to be considered as candidates.

6. Adjust the slider for Spelling Sensitivity from 80 (the default) to 70. Click OK to apply your preferences. ArcView uses the relaxed spelling sensitivity to find match candidates for the current address.

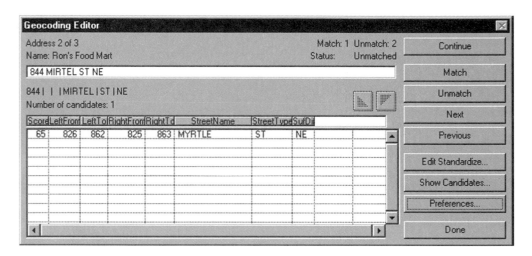

The address "844 MIRTEL ST" now has one candidate in the scrolling list. It has a match score of 65.

It's up to you whether to accept or reject the candidate. You decide that "MIRTEL" is probably an incorrect spelling of "MYRTLE," so you'll accept the highlighted candidate as a match.

If you know that an address in the address table is misspelled, you can edit it directly in the text box at the top of the Geocoding Editor. For example, you could change "844 MIRTEL ST" to "844 MYRTLE ST." Then there would be no need to adjust the spelling sensitivity. ArcView finds and scores all candidates using the new spelling. (Edits made to addresses in the Geocoding Editor are saved to the Av_add field of the geocoded theme table.)

7. Click the Match button to match the address to the highlighted candidate. The final unmatched address displays.

ENVIRONMENTAL SYSTEMS RESEARCH INSTITUTE, INC.

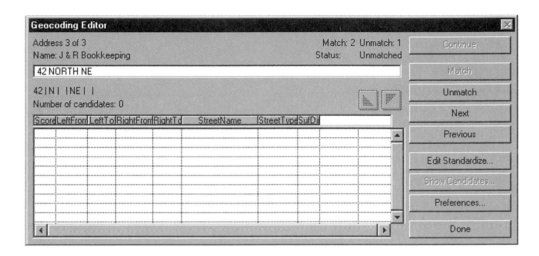

This unmatched address, "42 NORTH NE," has no candidates even though the spelling sensitivity has been relaxed. There appear to be no spelling errors, but if you look at the standardized version of the address, you'll see that the street name "NORTH" has become "N." It looks like there is a standardization problem, so you'll edit the standardization for this address.

Understanding standardization. ArcView uses a set of rules, called *standardization rules,* to determine which part of an address in the address table is the street name, which is the street type, and so on. Occasionally an address may be in an unusual format or may have too few or too many components. For example, a street name in a different language, such as "Avenida de las Estrellas," may have its address components in reverse order. Under ArcView's default standardization rules, "Avenida" (which means "Avenue") would be interpreted as part of the street name. If "Avenida" belongs to the street type field in the reference theme table, you won't get a match, even though the street exists and there are no spelling errors. If ArcView parses an ambiguous address differently from the reference theme table, you'll need to adjust the standardization manually to get a match. For more information, search for this Help Topic: *Edit Standardization (Dialog box).*

8. Click the Edit Standardize button. The Edit Standardization dialog box displays.

ArcView has parsed the address "42 NORTH NE" as follows: The street or house number (HouseNum) is 42, the prefix direction (PreDir) is N, the prefix street type (PreType) is not defined, the street name (StreetName) is NE, and the suffix street type (StreetType) and suffix direction (SufDir) are not defined.

Under its default standardization rules, ArcView was looking for the street name immediately before the street type. If the street name ("NORTH") had been followed by a street type (for example, "AVE"), ArcView would have standardized the address correctly. Because the street type was missing, ArcView interpreted "NORTH" as a direction instead of a street name. You'll change the standardization to make "NORTH" the street name and "NE" the suffix direction.

9. Click on the PreDir field to highlight it. Backspace to remove the **N** and press Enter to leave the field blank. In the StreetName field, type **NORTH** and press Enter. In the SufDir field, type **NE** and press Enter. The Edit Standardization dialog box now looks like this:

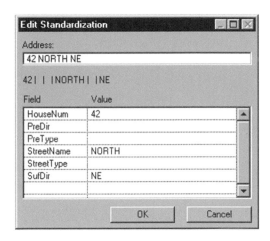

10. Click OK. The Geocoding Editor reappears. It now shows one candidate, with a match score of 75.

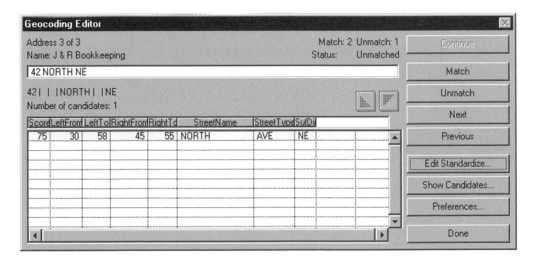

11. Click Match to accept the candidate, then click Done. The Re-match Addresses dialog box displays.

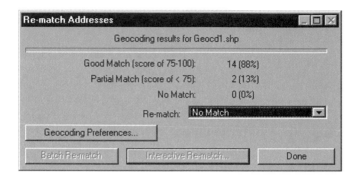

The re-match results now show that fourteen of the sixteen addresses in the customer list are good matches, with a score of 75 or better. Two addresses are partial matches, with a score below 75. There are no longer any unmatched addresses.

12. In the Re-match Addresses dialog box, click Done.

In the view, you can see that there are now sixteen customer locations in the Geocd1.shp geocoded theme.

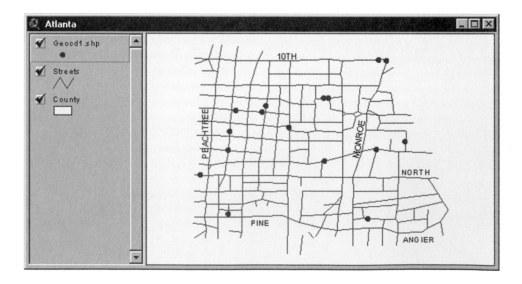

You've matched every address in the customer test list to the street theme, and you can see the distribution of addresses in the view. Now you could load the street theme for your entire area, make it matchable, and geocode a complete list of customers. Then you could use ArcView to create a new theme (see chapter 23) of sales territories on top of the geocoded theme.

Finding an address interactively

Suppose you're only interested in the location of one address. If you have a matchable reference theme, you can bypass the matching process and use the Locate Address button to find the address on the map.

13. Make the Streets theme active. Click the Locate Address button to display the Locate dialog box. (This button is available only when the active theme is matchable.)

14. In the Locate dialog box, type **400 4TH ST NE** (this dialog box is not case-sensitive), then click OK.

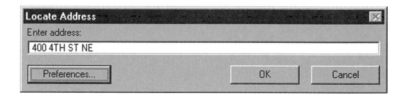

ArcView searches the street theme, finds the address, and places a black dot at the location in the view. If the location is outside the current display area, ArcView pans the view.

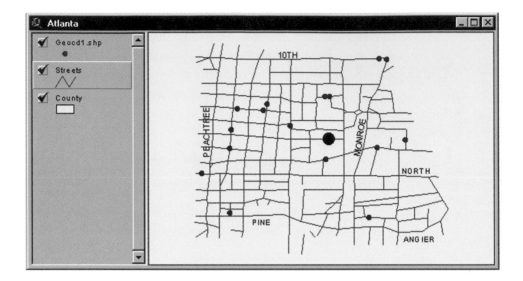

The dot ArcView draws when you use the Locate Address button is a graphic in the view, not a point feature in a theme (so it can't be used for spatial analysis). You can select and delete it with the Pointer tool. (If you wanted to save the location as a point shape in a geocoded theme, you could select the dot, choose Copy Graphics from the Edit menu, make the geocoded theme editable, and choose Paste from the Edit menu.)

In this chapter, you've seen how to create a geocoded theme of point locations using a reference theme and an address table. You made a street theme matchable, batch matched addresses in a table, interactively re-matched addresses that couldn't be matched by ArcView, and located a single address. Now you're ready to address geocode Ellsworth's entire customer list with its thousands of addresses.

If you want to go on to the next chapter, leave ArcView running. Otherwise, choose Exit from the File menu.

SECTION 8

Customizing ArcView GIS

This section introduces you to Avenue, the programming language that comes with ArcView GIS. You'll see how to use Avenue to create a custom application, including writing Avenue scripts and customizing ArcView's interface. Then you'll test the new interface.

SECTION 8:
Customizing ArcView GIS

Introducing Avenue

How Avenue and ArcView GIS work together

Creating the custom application

Writing Avenue scripts

Customizing the ArcView GIS interface

Testing the new interface

CHAPTER 27

Introducing Avenue

ArcView GIS meets a wide variety of GIS needs. But you might still need to customize it. And you can, with ArcView's programming language called *Avenue*.

You can use Avenue™ software to customize the standard graphical user interface (GUI) that comes with ArcView. For example, you can reorganize the controls (menus, buttons, tools, and popups), change text or icons, and add or remove choices. With Avenue you can also create new functions for a specific application. For example, you can combine a series of steps you frequently perform and execute them with a single click of a button. You can even use Avenue to develop a complete application that has its own GUI.

Avenue gives you an easy-to-use framework for customizing controls and creating new functions. You use the Customize dialog box to modify controls and the Script Editor to write Avenue programs, called *scripts*. Your scripts contain the code that executes a new function. By using controls and scripts together, you build a new ArcView GUI.

ENVIRONMENTAL SYSTEMS RESEARCH INSTITUTE, INC.

How Avenue and ArcView GIS work together

Avenue is an object-oriented programming language. An *object* is an element, such as a view, theme, button, or symbol, that you work with in ArcView. Objects with common characteristics belong to the same *class*. Each object is associated with a set of actions or *requests*. For example, requests for a layout object include opening, closing, and printing. A marker symbol object has a different set of requests, such as setting marker size and color. If you can identify objects and their associated requests, you can write Avenue scripts.

Avenue and ArcView use the same interface; in fact, all menus, buttons, tools, and popups (collectively called "controls") in the ArcView interface run Avenue scripts. When you add a new view, you're actually running the Avenue script "View.Add." To change the way a control works, you modify the script it runs.

You create an Avenue script by opening a Script document window in ArcView, then using the Script Editor to write, compile, run, and debug it. In this chapter, you'll see how Avenue might be used to customize the ArcView interface for a specific application. You won't be able to perform the steps yourself, because the Script Editor is not available in the sample version of ArcView that accompanies this book; however, the finished project is provided so you can run the application.

A standard version of ArcView includes Avenue and all of the customization capabilities mentioned here.

Creating the custom application

Maria works as a public information consultant. She's been hired to design a public information system for the Visitor's Bureau in Atlanta, Georgia. Her goal is to make it easy for visitors to find their way around the city. She plans to use ArcView to set up automated information kiosks at several sites for the public to view maps showing locations of points of interest, shopping centers, hotels, and banks. She needs a simple, easy-to-use interface anyone can understand. She doesn't want visitors to delete or modify her project, so she plans to remove all existing controls, then add new controls that visitors can use to quickly switch between the various map displays. Visitors will be using a sophisticated GIS without even knowing it.

Maria creates an ArcView project containing a number of different views, one for hotels, one for shopping centers, and so forth. She uses the standard ArcView interface to create the project components, so she has full processing capabilities. She hides the Table of Contents in each view so that it can't be changed.

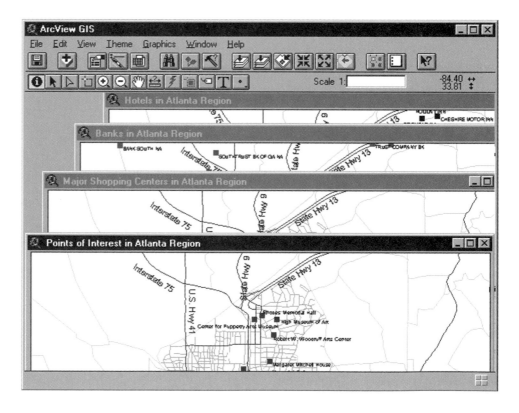

Maria decides to have only five buttons on her new View interface: one to display each view, and one Help button. To implement her design, she'll write five Avenue scripts, four to pop up the different views, and one to pop up a Help message. She'll also remove all other tools, buttons, and menus from the View interface. She'll use the menu bar to label the function of each button.

ENVIRONMENTAL SYSTEMS RESEARCH INSTITUTE, INC.

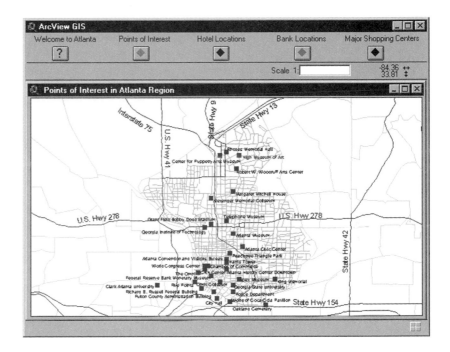

Writing Avenue scripts

Maria creates a simple Avenue script that finds a view, then makes it active, bringing it to the front of the screen. From the ArcView Project window, she opens a new Script window and types in the script.

```
Script1                                                    _ □ ×
'Points of Interest

'The next line of code finds the "Points of Interest" view:
myView = av.GetProject.FindDoc ("Points of Interest in Atlanta Region")

'The next line of code makes the "Points of Interest" view
'the active document:
myView.GetWin.Open
```

Understanding Avenue syntax. Scripts contain action lines (perform an action) and comment lines (explain what the script is doing). The format of an action line is: *Object.Request.* Comment lines begin with an apostrophe (').

Maria uses Script Properties from the Script menu to name the script and enter a comment describing what it does.

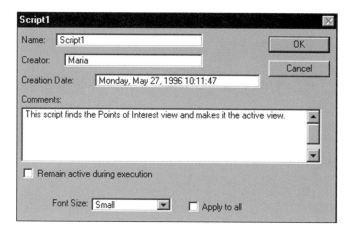

She creates three more scripts like this one. Each finds a different view and is named accordingly.

She writes a fifth script, called "HelpBox." It opens a message box that provides instructions for the user.

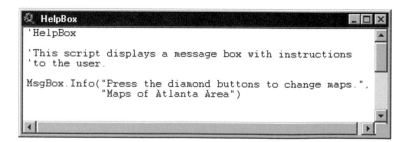

She compiles each script using the Compile button on the Script button bar. Then she runs each script using the Run button to make sure it works. She verifies that each view displays at the front of the screen. When she runs the "HelpBox" script, a help message box displays.

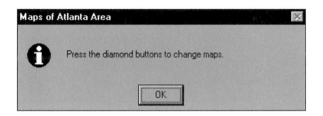

Customizing the ArcView GIS interface

Maria's final task is to customize the ArcView GIS interface. She'll remove all the existing controls, then add her own and link them to the five scripts she's written. She uses the Customize dialog box, which she accesses by double-clicking on any blank portion of the Script button or tool bar.

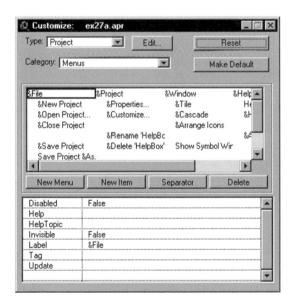

The Type field lists the parts of the ArcView interface you can customize (views, tables, charts, layouts, scripts, projects). The Category field lists the kinds of controls you can customize (menus, buttons, tools, popups). Maria wants to customize the View menu bar. By pressing the Delete button repeatedly, she removes all of the View menus.

She changes the Category to Buttons and deletes all the buttons, then changes the Category to Tools and deletes all the tools. The View interface is now completely blank.

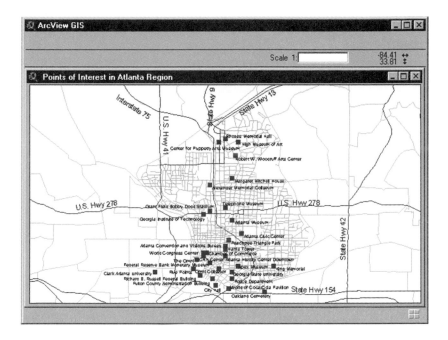

With the "Buttons" category selected, Maria presses New to add a new button to the Button bar. By default, no icon appears on the button.

Next she sets the button's Click property. This property names the script to execute when the user clicks the button. Double-clicking on the "Click" property field displays the Script Manager, a dialog box that lists all scripts available in the project. Maria selects the "Points of Interest" script.

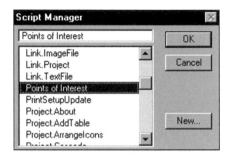

This script is now assigned to the new button.

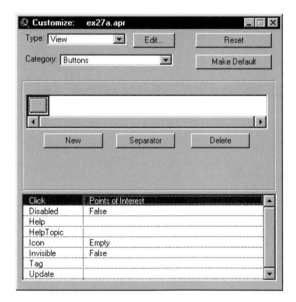

ENVIRONMENTAL SYSTEMS RESEARCH INSTITUTE, INC.

Next she double-clicks on the "Help" field to define the help message the user sees in the status bar when placing the cursor over the button.

The button needs an icon, so Maria double-clicks on the "Icon" field to display the Icon Manager. She chooses the red diamond icon for this button.

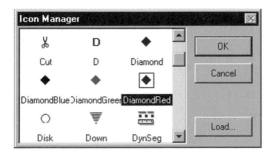

The icon displays in the dialog box.

The first button is now defined. Maria creates the other buttons, assigning a script, help string, and icon to each. By dragging the buttons, she can change their order.

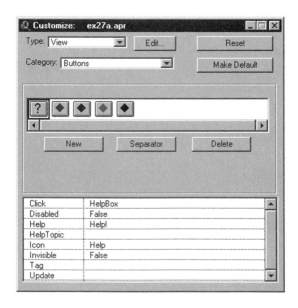

She uses the Separator button to put some space between the buttons.

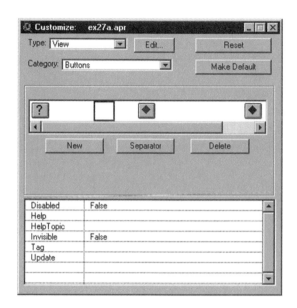

For menus, Maria creates text labels to describe each button. With "Menus" selected in the Category list, she chooses New Menu, then double-clicks on the "Label Property" field. Here she specifies the menu text that will appear in the Menu bar above each button.

The View interface is now complete, but there's one more thing to do. With the Project window active, Maria chooses Properties from the Project menu. She specifies "Points of Interest" as the Startup script (the script that executes when you start up the project), then clicks OK. When the project opens, it will automatically display the Points of Interest view.

Testing the new interface

You can open Maria's project and test each button to see that everything works. With just a few simple scripts and some customizing, ArcView is transformed into a point-and-click map display interface that doesn't require any GIS knowledge.

The application you're about to see uses these display options: 1024x768 screen resolution, 256 colors, small fonts. If your settings are different, you should change them to these for best viewing results.

Exercise 27a

1. If necessary, start ArcView. From the File menu, choose Open Exercise. In the Exercises scrolling list, select "ex27a," then click OK. When the project opens, you see the modified View interface and the Points of Interest view.

ENVIRONMENTAL SYSTEMS RESEARCH INSTITUTE, INC.

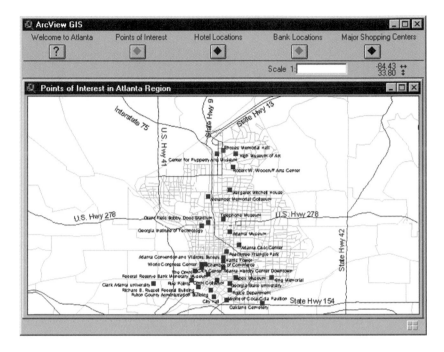

2. Click the Help button, labeled "Welcome to Atlanta." A message box with instructions displays. Click OK to dismiss it.

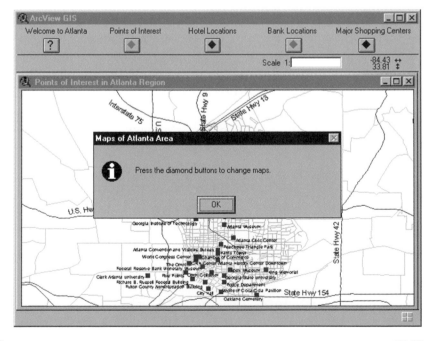

3. Click each of the other buttons to change the map display.

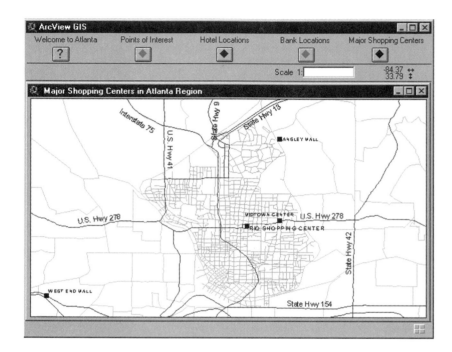

4. When you finish, close all views. You see the Project window. Choose Exit from the File menu to exit ArcView.

Studying existing Avenue scripts is one way to learn Avenue. In the standard ArcView GIS release, Avenue Help comes with a Script Library — a collection of scripts useful for many applications. You can copy any of these scripts into the Script window and modify them to suit your needs. All of the system scripts that come with ArcView are also available to you.

You've seen how flexible the ArcView interface is and how, with a little Avenue programming and customization, you can extend it far beyond its original design. This is only the beginning. The possibilities are endless.

ENVIRONMENTAL SYSTEMS RESEARCH INSTITUTE, INC.

SECTION 9

ArcView GIS extension products

This section introduces you to two ArcView GIS extension products. ArcView extensions are software products you purchase to give ArcView expanded capabilities. In chapter 28, you'll see how the ArcView Network Analyst extension gives ArcView the ability to find the best route to reach many stops, find the closest facility to an event, and find the service area around a site. In chapter 29, you'll see how the ArcView Spatial Analyst extension gives ArcView the ability to create, analyze, and overlay grid themes. New extensions for ArcView are being developed all the time. ArcView Internet Map Server and ArcView Dialog Designer will be available soon. Visit ESRI's Web site at www.esri.com for information on new extensions.

SECTION 9:
ArcView GIS extension products

Introducing ArcView Network Analyst

Preparing to use ArcView Network Analyst

Finding the best route

Finding the closest facility

Finding service areas

CHAPTER 28

Introducing ArcView Network Analyst

A *network* is any set of interconnected linear features, like roads, pipelines, or airline routes. Networks enable you to get where you're going, to communicate with others, and to deliver essential resources, such as gas, water, and electricity. The ArcView Network Analyst extension helps you map and analyze these linear features.

You can solve three kinds of problems with ArcView Network Analyst. You can find the *best route* along stops in a network (for example, the shortest package delivery route for an express mail service). You can find the *closest facility* to a location (for example, the nearest tow truck company to a stranded motorist). You can also find a facility's *service area* (for example, the area around a pizza parlor that can be reached within thirty minutes). The solution to each network problem is stored as a theme in ArcView's shapefile format, so it can be updated when the network changes.

Network Analyst can also create a written set of street directions telling you how to follow a route in a network. The directions can include local landmarks that appear along a route. You can print or save the directions in a file.

Preparing to use ArcView Network Analyst

Before you can use ArcView Network Analyst, you need to load it by choosing Extensions from the File menu, then checking the box next to Network Analyst.

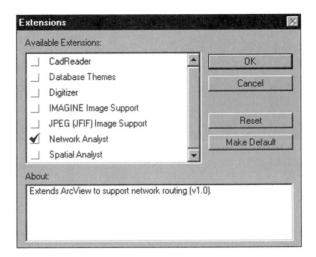

Network Analyst is an extension to ArcView GIS that you purchase separately. It isn't part of the demonstration copy of ArcView that comes with this book, so you won't be able to load it or perform the steps in this chapter yourself.

You also need a matchable street theme. You make a street theme matchable by choosing its address style in Theme Properties. (For details on making a theme matchable, see chapter 26.) If necessary, you can add a field to the attribute table of the street theme to indicate closed streets and the flow direction of one-way streets. You can also create drive time fields that show the time it takes to travel a street. You might want to have many drive time fields indicating normal and rush hour driving times.

Ron's at his desk in the Traffic Planning Department at City Hall when a big earthquake rolls through the area. After picking up the objects strewn all over his office and calming his nerves, he joins the rest of the staff as they implement emergency procedures. One of the tasks assigned to his department is to assess the damage to streets and highways. This information is vital to police, fire, and ambulance services.

Under normal circumstances, Ron studies traffic flow patterns using a matchable theme of city streets. Now that theme will serve as an accessibility map for emergency workers. Ron sets up an emergency command post where he receives radio, telephone, and television reports about fallen overpasses, streets flooded by ruptured water lines, and roads blocked by debris. He keeps track of this information in the attribute table of the City Streets theme.

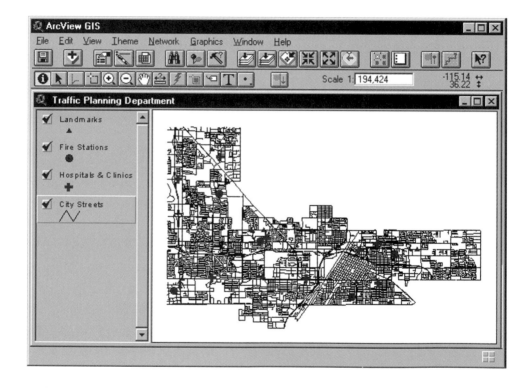

Because Ron has the Network Analyst extension loaded, there's an extra menu item, called Network, in the ArcView interface. There are also two extra buttons and an extra tool created by Network Analyst.

ENVIRONMENTAL SYSTEMS RESEARCH INSTITUTE, INC.

As each damage report comes in, Ron adds the information to the attribute table for the City Streets theme. He changes the values in the Oneway field to indicate which streets have become one-way or completely impassable.

St_prefix	St_name	St_type	Average_sp	Number_lan	Oneway	Drivetime
N	DECATUR	BLVD	50	2	FT	0.09
	SAWYER	AVE	25	2		0.03
	COMSTOCK	DR	25	2		0.11
	US-95		55	3	TF	0.27
	PRISINZANO	CIR	5	2		0.15
	EDWARD	AVE	25	2	N	0.20
	MADELINE	DR	25	2	N	0.14
	SAN SIMEON	ST	25	2		0.05
	VALLEY	DR	25	2	N	0.13
	COMSTOCK	DR	25	2		0.06
W	LAKE MEAD	BLVD	50	4	N	0.11

If the street is passable in one direction only, Ron changes the value in the Oneway field to TF or FT (To-From or From-To); he changes the value to N (Not passable) if the road has been closed. (Each street has a From-To and a To-From direction. The From-To direction is the direction you travel if you begin at the starting point of a street—the first point of the street segment that was drawn when the street data was created—and move toward the ending point. The To-From direction is the reverse.)

Ron has recorded, in the Drivetime field, the average time in minutes it takes to travel each road under normal circumstances. As he gets reports about traffic congestion from police and emergency workers, he selects the congested streets in the attribute table and uses the Field Calculator (see chapter 15) to multiply the Drivetime field by a factor of three or four or more, depending on his estimate of the congestion.

Having updated the attributes in the theme table with the latest street conditions, Ron's ready to use Network Analyst to find unblocked routes and keep emergency services flowing to the places they're needed.

Finding the best route

After the quake, truckloads of emergency medical supplies are arriving from all over the country. Ron and his coworkers must route the supply trucks around blocked streets and highways to functioning hospitals and clinics. Ron uses Network Analyst to find the fastest routes and create directions for the truck drivers.

Ron highlights the damaged and closed streets in the view. He can use the Query Builder (see chapters 13 and 14) to select closed streets. Network Analyst considers a street closed if it's highlighted or has an "N" in the attribute table's Oneway field.

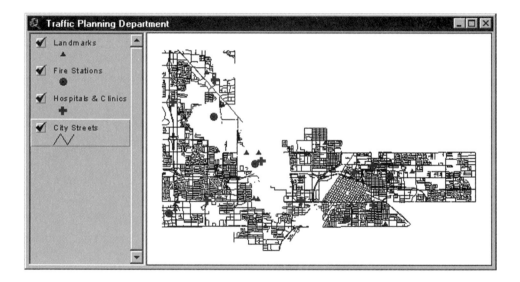

Ron wants to find a route through the City Streets theme that most efficiently reaches each point in the Hospitals and Clinics theme. First, he needs to define the problem for Network Analyst. He chooses Find Best Route from the Network menu.

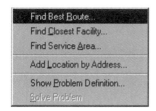

The Route1 dialog box displays. This is where Ron defines the stops (the hospitals and clinics) along the route.

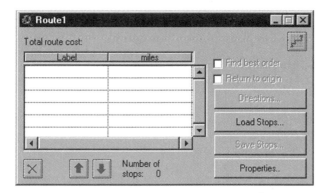

  Ron can define the stops in three ways. He can click on locations in the view with the Add Location tool. He can also type in street addresses by clicking the Add Location by Address button. In this case, he presses the Load Stops button in the Route1 dialog box to load the stops from the Hospitals & Clinics point theme.

The Load Stops dialog box appears and Ron selects the Hospitals & Clinics theme. This will include every point feature in the theme as a stop.

Now each hospital and clinic is listed as a stop along the route in the Label column. The names are taken from the Label attribute in the Hospitals & Clinics attribute table.

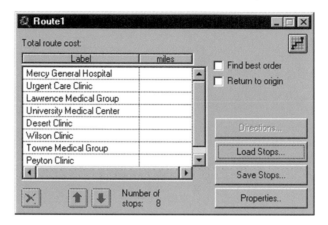

In the view, each point in the Hospitals & Clinics theme has a green square over it, indicating that it's a potential stopping point. (Notice that one potential stop is in the highlighted zone of closed streets and therefore unreachable by truck.) There's also a new, empty theme, called Route1, in the view. This is where Network Analyst will display the best route.

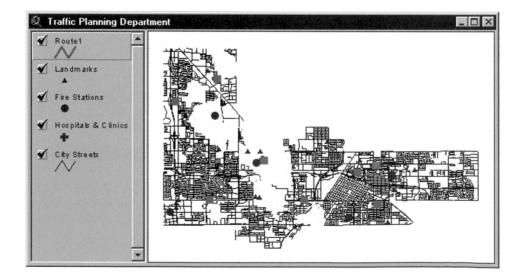

Before creating the best route, Ron clicks the Properties button in the Route1 dialog box. The Properties dialog box appears.

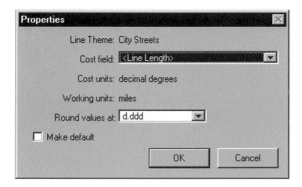

Network Analyst determines the best route by finding the route with the lowest "cost." Cost can be defined as distance, travel time, or some other attribute. By default, a route's cost is found by adding the lengths of all line segments in the route. Network Analyst performs this calculation for all possible routes that pass each stop and chooses the route with the least cost. This is the shortest route.

Under normal circumstances, finding the shortest route is fine, but with the traffic congestion caused by closed freeways and streets, the shortest route may take longer to drive. The best route in this case is not the shortest, but the fastest. Ron changes the Cost field to Drivetime. Network Analyst will find the route with the lowest cost in terms of time rather than distance.

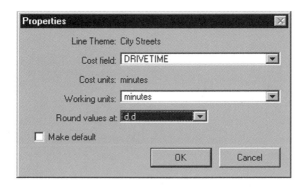

The Route1 dialog box reflects this change. The cost column is labeled "minutes" instead of "miles."

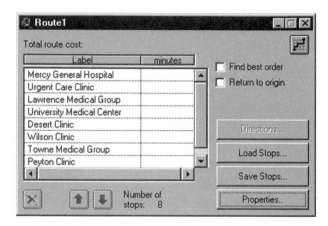

 Ron has provided all the necessary information, so he clicks the Solve button. Network Analyst finds the fastest route to each hospital and clinic. It displays the time to drive the entire route in the Total route cost field at the top of the dialog box. The minutes column shows the elapsed time from the route's starting point to each hospital or clinic. ArcView tells you that Desert Clinic (the one located in the section of closed streets) couldn't be reached. This clinic will need its supplies delivered by helicopter.

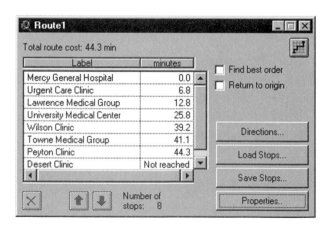

The route also displays in the view. Notice that it avoids the damaged streets. This route may not be the shortest in terms of distance, but it has the shortest drive time.

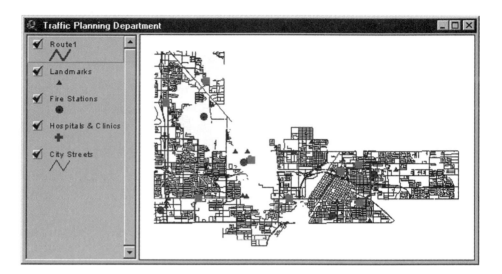

Ron can label the street names and print a copy of the map for each supply truck driver. Because many drivers are unfamiliar with the area, he'll also use Network Analyst to print a set of directions. From the Route1 dialog box, Ron clicks the Directions button. The Directions dialog box displays, giving the directions to each stop.

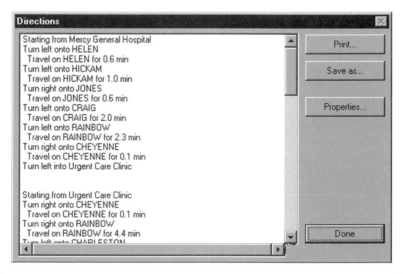

Ron decides to customize the directions to make them clearer. Although it made sense to calculate the route based on drive time, it will be better to give directions in terms of mileage, so the truck drivers can check their odometers as they drive. He clicks the Properties button to display the Direction Properties dialog box.

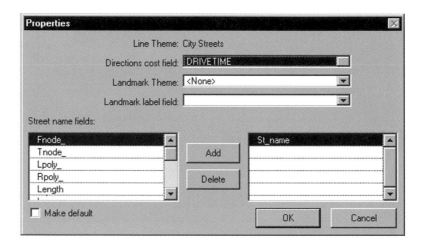

Ron changes the Directions cost field to <Line Length>. The directions will now show the distance in miles, even though the route was calculated in minutes. To make the directions even clearer, Ron adds landmarks by selecting a theme in the Landmarks Theme field. (The Landmarks theme contains the names and street addresses of points of interest in the city.) He also wants to add street directions, such as N or S, and street suffixes, such as Ave or Blvd, to the street names. He selects the appropriate fields from the Street name fields scrolling list and adds them to the scrolling list on the right.

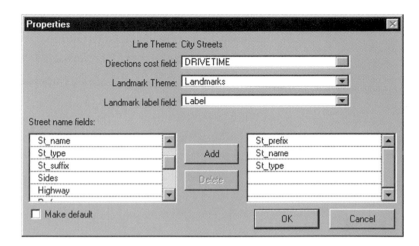

When Ron clicks OK, the directions are customized to suit his purpose.

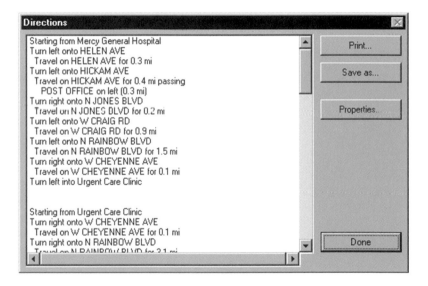

Now the directions include information about landmarks along the route, like the post office 0.3 miles along Hickam Avenue on the right side.

Ron gives the map and directions to each truck driver. As traffic conditions change in the days ahead, Ron will update the City Streets attribute table. He'll choose Show Problem Definition from the Network menu to redisplay the Route1 dialog box and click the Solve button to update the map and directions. With Network Analyst, Ron can ensure that the truck drivers always use the best available routes.

Finding the closest facility

The Fire Department has been called to a major blaze in the northeast part of town, but the nearest fire station is damaged and not operational. Because of road closures, the next-closest station may or may not be able to get a fire engine to the scene. Ron has the most up-to-date road closure information stored in ArcView, so he uses Network Analyst to determine which station should respond. He also prints a set of directions to guide the engines along passable roads.

To solve this network problem, Ron selects Find Closest Facility from the Network menu. An empty theme called Fac1 (Facilities theme 1) is created in the view. This theme will contain Network Analyst's solution to the problem Ron defines.

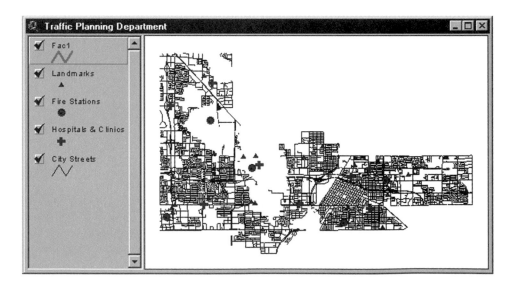

ENVIRONMENTAL SYSTEMS RESEARCH INSTITUTE, INC.

In the Fac1 dialog box, Ron sets Facilities to the Fire Stations theme, and 1 as the Number of facilities to find. He chooses the Travel to event option because the fire engines will travel from the station to the fire. He could set a cutoff cost to select only routes that have a lower cost than the one he specifies, but under the circumstances, he doesn't want to eliminate any possibilities. Ron has set the properties (by clicking the Properties button) to show the route cost in minutes instead of miles.

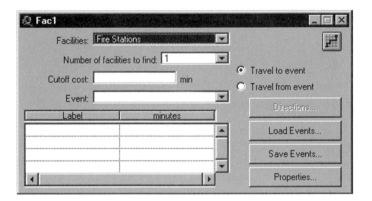

 Now Ron needs to enter the location of the fire in the Event field. Since there's only one event to locate, Ron clicks the Add Location by Address button in the view button bar. When the Locate dialog box appears, he enters the fire's location.

In the Fac1 dialog box, the Event box now shows the fire's location. Ron has completely defined the network problem, so he clicks the Solve button. Network Analyst calculates which fire station will be able to respond most quickly. It displays the name of the closest station in the Label field and the estimated travel time in the minutes field.

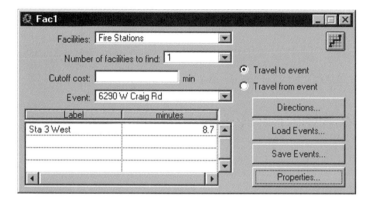

The view shows the fastest route to the fire from the selected station. It's not the closest station, but it's the one that can respond most quickly under the given conditions.

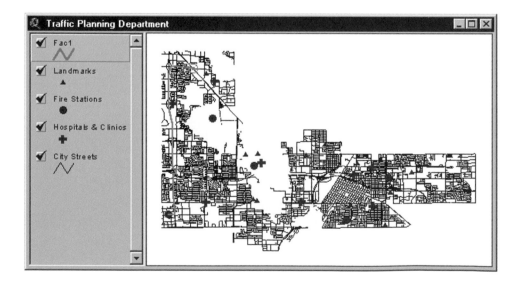

Just as he did for the hospital route, Ron creates directions and customizes them to show landmarks, full street names, and distances in miles instead of minutes.

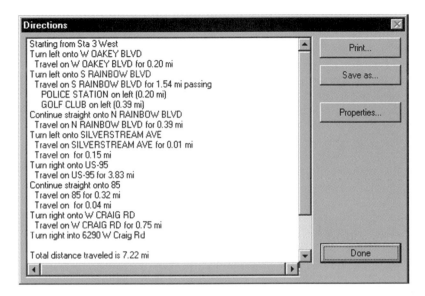

The engines are on their way with directions that guide them around the traffic congestion to the fire.

Finding service areas

Ron and his coworkers have been asked to identify those areas where ambulances can't reach people within ten minutes. Temporary tent clinics will be set up in these areas and helicopters used to transport seriously injured people to hospitals. Ron will use Network Analyst to determine the areas that are reachable over open roads within ten minutes of each hospital and clinic. He'll also produce a map showing the less accessible areas that need medical services.

To solve this network problem, Ron chooses Find Service Area from the Network menu. Two empty themes are created in the view, Snet1 and Sarea1. These themes will contain the solution to the service area network problem. Snet1, a line theme, will show all city streets within ten minutes of a hospital or clinic. Sarea1, a polygon theme, will show polygons that cover the ten-minute drive time area around each hospital. There will be one polygon for each hospital or clinic and it will encompass the area that's reachable within ten minutes.

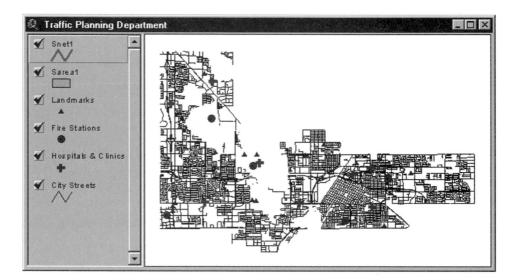

The Sarea1 and Snet1 dialog box also displays. Ron loads the Hospitals & Clinics theme by clicking the Load Sites button. He sets the properties to display minutes instead of miles, then types 10 in the minutes field for each hospital or clinic. This is the maximum drive time an ambulance can take to reach an injured person.

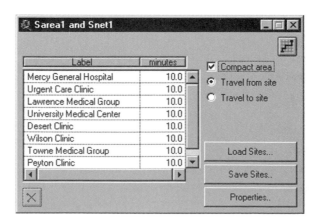

Ron also checks the Compact area check box. If this box isn't checked, Network Analyst will create smooth polygon shapes for the service area to make the map look better for presentation. If Compact area is checked, Network Analyst draws the ten-minute service area precisely, even though the polygon outline will be jagged. Ron doesn't care how the map looks; he wants the outlines of the ten-minute service area to be exact.

 After Ron clicks the Solve button, Network Analyst creates lines in the Snet1 theme showing all the streets within ten minutes of medical help. Network Analyst also creates polygons in the Sarea1 theme showing the ten-minute zone around each hospital and clinic. Areas not included in these zones will need temporary clinics and helicopter service.

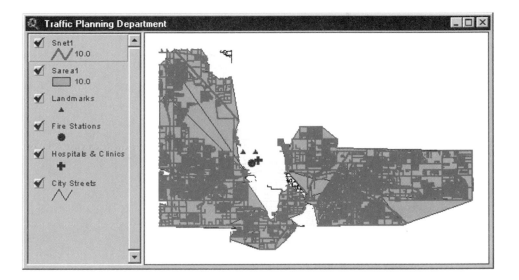

Ron can print these maps and give them to the planners in charge of getting medical service to the entire city.

As repair crews reopen streets, Ron can easily update any of the maps and directions he's created. He changes the street conditions in the attribute table, redisplays the problem definition, and creates a new solution.

Using ArcView Network Analyst, Ron can respond to rapidly changing conditions and help keep critical services going where they're needed.

ENVIRONMENTAL SYSTEMS RESEARCH INSTITUTE, INC.

CHAPTER

29

SECTION 9:
ArcView GIS extension products

Introducing ArcView Spatial Analyst

The advantages of grids

Creating grids from features

Creating grids from elevation data

Doing math with grids

Building spatial models

Introducing ArcView Spatial Analyst

You're already familiar with spatial analysis operations like theme-on-theme selection and spatial join. ArcView Spatial Analyst, an extension purchased separately, gives you capabilities far beyond these. With Spatial Analyst, you can map the distribution of data such as population density, elevation, and distance. You can overlay multiple themes, as if each theme were drawn on a separate transparency, to analyze the relationships among them. You can build complex spatial models to help you make decisions involving many variables. To appreciate the power of Spatial Analyst, you need to understand how it represents map features as *grids*.

There are basically two ways to represent map features in a GIS. One is to depict them as points, lines, and polygons. The other is to assign them numeric values in a grid of equally sized cells. Points, lines, and polygons are organized in feature themes. Cell-based map features are organized in grid themes.

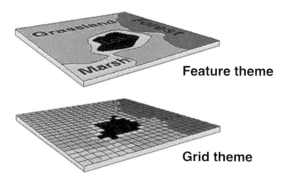

Feature theme

Grid theme

In feature themes, map features are represented as geometric shapes. In grid themes, map features are represented as cells that share a numeric value. A lake can be represented by a polygon of a certain size and shape, or it can be represented by a cluster of cells that have the same value.

Spatial Analyst can convert any feature theme to a grid theme, so you can use it with all your ArcView data. Similarly, you can convert any grid theme to a feature theme. Throughout this book, you've seen what can be done with feature themes. In this chapter, you'll see what can be done with grids.

The advantages of grids

Grid themes can represent the same geographic objects as feature themes (objects like lakes, highways, and hotels). They can also represent *continuous surfaces*. A continuous surface, like elevation or temperature, is a geographic feature or phenomenon that lacks definite boundaries and tends to change gradually. You can't measure a continuous surface at every single point in a location, but you can take measurements at several sample points. Spatial Analyst uses these sample points to create a grid theme of estimated values for every cell in the location. In this example, sample points have been used to create a grid theme of elevation, shown as a relief map.

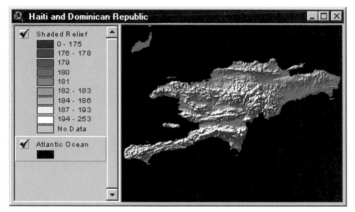

Continuously changing elevation values are shown as graduated cell colors.

Because grid themes are essentially rows and columns of numbers, you can do a wide range of mathematical operations on cells. For instance, you might add a certain value to each cell in a temperature grid to map a predicted heat wave. In addition, you can take multiple grid themes, align them with one another in a process called *spatial overlay*, and add, subtract, or combine their cell values in a variety of ways. Later in the chapter, you'll see how spatial overlay lets you build models to solve complex spatial problems.

ENVIRONMENTAL SYSTEMS RESEARCH INSTITUTE, INC.

Creating grids from features

Spatial Analyst can convert feature themes directly to grid themes, or it can work with feature and grid themes together. As you'll see, this makes it possible to do some interesting and useful studies.

Creating a surface grid from sample points

Chris is a farmer who wants to reduce the cost of fertilizing his fields. First, he measures soil nutrients at a number of sample points. From his point theme of sample data, Spatial Analyst then generates a surface map of estimated nutrient levels across the entire farm. Since Chris knows the optimum level, he can create a grid of fertilizer requirements by subtracting actual from ideal values. While he's at it, he draws a 300-meter buffer zone around the stream to avoid polluting the water. Chris saves money by applying fertilizer intelligently and economically.

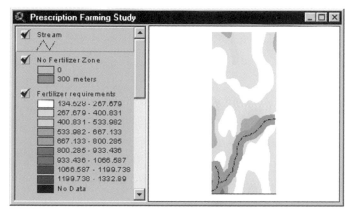

From a point theme of soil samples (not shown), Spatial Analyst creates a continuous surface grid. Chris uses this grid to make a map of fertilizer requirements. He adds another grid theme that shows a 300-meter buffer around the stream.

Creating a distance grid from polygons

Michelle is on a committee studying noise levels associated with an airport expansion. She uses Spatial Analyst to create a grid theme that measures the distance from each cell to the expansion. (Each cell's value is its distance from the nearest edge of the polygon.) The distance grid can then be overlaid with a noise decay grid to show which city residents will be most affected.

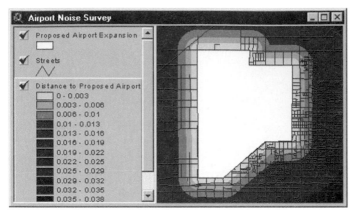

Michelle uses a polygon theme (Airport Expansion) to create a grid theme of distance. In the grid theme, the area around the airport is divided into cells. Each cell's value is its distance from the airport. The cells grade from orange to violet as distance from the airport increases.

Defining areas nearest to points

Rob owns a movie theater chain. Each theater manager is responsible for distributing flyers and coupons to the neighborhoods within his or her customer territory. To determine each manager's territory, Rob creates a proximity grid theme. Spatial Analyst measures each cell's distance to every theater, and assigns it to the nearest one.

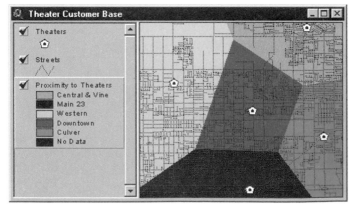

Rob uses a point theme (Theaters) to create a proximity grid theme showing the area served by each theater. Each cell in the proximity grid theme receives a value according to which theater is nearest to it. Cells with the same value (and color) are nearest the same theater.

ENVIRONMENTAL SYSTEMS RESEARCH INSTITUTE, INC.

Distributing values around points

Regina is a planner for a health care provider. She's researching locations for a new urgent care clinic. She uses a point theme of cities (with a population attribute) to create a population density map for the entire area. Regina can see where the population density is greatest and use that as a factor in her evaluation.

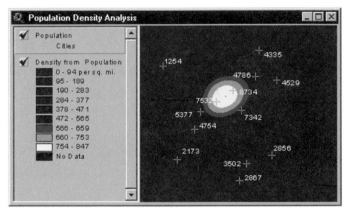

From a point theme of cities, Regina distributes population values according to a formula that takes into account the locations of nearby points. The result is a grid theme of population density.

Creating grids from elevation data

You can create grid themes from feature themes, images, other grids, or imported data. Elevation data, available from government sources, can be imported into grid format by Spatial Analyst. Elevation data can be used to create surfaces that resemble actual terrain and to derive information that's related to elevation.

Creating contour, hillshade, and visibility maps

Gary, a geologist, wants to create maps of Mount Saint Helens to show how rock was redistributed by the eruption. From a Mount Saint Helens digital elevation model (a file of sample elevation values), he uses Spatial Analyst to generate elevation, contour, and hillshade maps. The hillshade map simulates a three-dimensional image. Gary wants to take aerial photographs, so he draws a proposed flight path over the mountain. Spatial

Analyst creates a visibility map that shows the areas that can be seen from a given point by a plane flying at 3,500 feet.

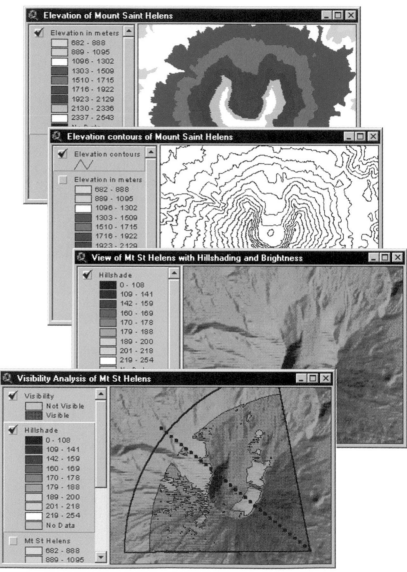

Gary creates an elevation grid (top) of Mount Saint Helens from an imported elevation file. He uses the elevation grid to generate a contour map (second from top). Each line in the contour map marks a 100-meter change in elevation. The elevation grid is also used to make a hillshade grid (second from bottom). Finally, he creates a visibility map (bottom) based on the height of the plane and the camera's view angle.

Creating slope and aspect maps

Lisa is a botanist studying plant species in the Grand Canyon. The native vegetation types have specific slope and sun requirements. From elevation data, Spatial Analyst can create slope and aspect maps. Slope measures the steepness of terrain; aspect shows the compass orientation (north, south, and so on) of the slope. Lisa can predict where certain plant species will be found by looking at these maps.

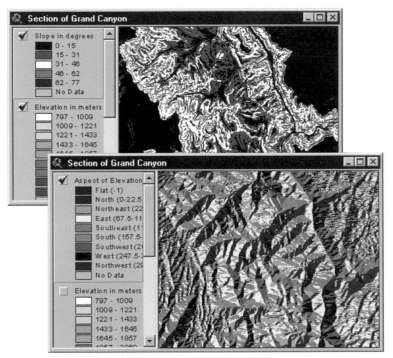

Lisa creates a slope grid (top) from an elevation model of the Grand Canyon. From the slope grid, she derives an aspect grid, showing the direction of slope.

Creating hydrology maps

Randy is a hydrologist who wants to study a potential contaminant spill. He uses an elevation grid to create a map of flow direction (like an aspect map, this measures the direction of downhill slope). He then chooses points on the elevation grid that represent hypothetical spill sources. Spatial Analyst traces the probable path the contaminant would follow.

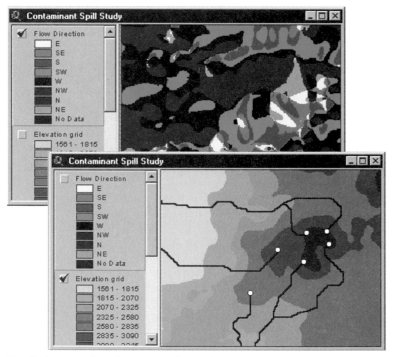

Randy derives a flow direction grid (top) from an elevation grid. He then marks hypothetical spill points, shown by white dots, on the elevation grid (bottom). Spatial Analyst uses the flow direction and elevation values to compute the contaminant's probable downhill path.

Doing math with grids

You can add, subtract, multiply, and divide the values in grid cells to create a new grid. You can also apply a variety of mathematical functions to perform highly sophisticated spatial analysis.

Changing values in grid cells

Melinda, a meteorologist, is modeling the behavior of tropical storms. One component in her model is a grid showing wind direction. Each cell in the wind direction grid has a numeric value from 0 to 360 degrees. By adding a constant value to each cell, Melinda can simulate a shift in the storm's direction. She uses the wind direction grids with other grids to make assumptions about where and when a storm might reach land.

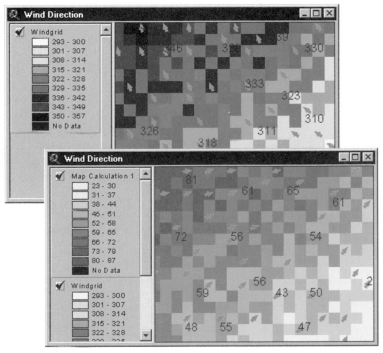

Using Spatial Analyst, Melinda adds 90 degrees to each cell value in the first grid to generate a second wind direction grid.

Creating statistical tables and charts from a grid

Paul is a sales manager for a company that supplies restaurants and fast food stores. He's decided to review his salesmen's territories to make sure they're equitable. In the Number of Restaurants grid theme, each cell's value is the number of restaurants it contains. Paul overlays a polygon theme of sales territories. Spatial Analyst uses each polygon to divide the grid theme into zones, then counts the number of restaurants in each zone. The results are stored in a table and charted, so Paul can see how many restaurants lie in each salesman's territory. (Many other statistics, such as minimum, maximum, and mean values, are also stored as fields in the table.)

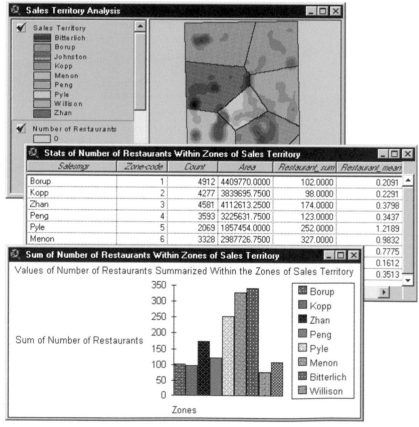

Paul uses polygons from the Sales Territory theme to count the number of restaurants within each territory. The numbers are stored in a table and charted.

ENVIRONMENTAL SYSTEMS RESEARCH INSTITUTE, INC.

Using a graphic to get grid statistics

Tina, a financial analyst for the Water Department, is studying the revenue impact of a proposed water tower. Using Spatial Analyst, she draws a circle around the area that will draw water from the tower. Because different rates are charged to agricultural, industrial, and residential customers, Tina needs a breakdown of land use by area within the circle.

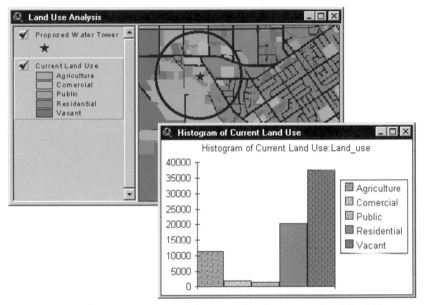

Tina uses Spatial Analyst to calculate the area of land use types that fall within a circle drawn on a land use grid.

Measuring variety

Jorge, a biologist, wants to measure the diversity of plant life in a region. He has a land cover grid, which shows the types of vegetation found in his study area. Using Spatial Analyst, he counts the number of different types of land cover surrounding each cell. Cells whose neighbor cells have many different values are part of a biologically diverse area.

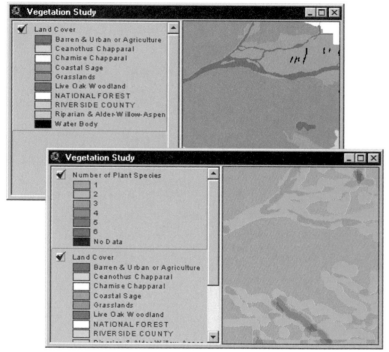

Spatial Analyst looks at the cells surrounding each cell in the Land Cover grid and counts the number of different values. Jorge creates a new grid in which each cell's value is the number of different neighboring land cover types.

Building spatial models

Spatial Analyst's ability to overlay grid and feature themes makes it a powerful tool for creating spatial models. A model is a scaled-down representation of conditions in the real world, and each layer of data in a model represents one of these conditions. With Spatial Analyst, you can integrate, weigh, and evaluate as many different layers of data as are appropriate.

Spatial models are used to simplify complex problems, like where to locate a factory, by breaking them down into conditions that can be represented on different grids (land cost, zoning codes, distance from distribution networks, and so on). Spatial modeling is also used to simulate processes like the spread of a fire or an oil slick. Starting conditions (for instance, temperature, wind strength, and dryness of vegetation) can be varied to predict possible outcomes.

Suppose Judy is looking for the best place in the valley to grow oranges. Oranges like sandy soil and dry climates. By overlaying a grid of soil type and a grid of climate, she can test each location to see if it meets these two conditions. (The two grids cover the same geographic area and have the same size and number of cells, so they can be lined up.) Each cell in the soil grid is compared with the corresponding cell in the climate grid. A logical test is applied (is the soil sandy and the climate dry?). Locations that are both sandy and dry pass her test; all others fail.

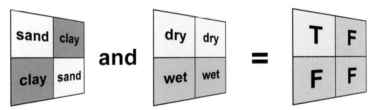

Spatial Analyst compares the corresponding cells in the two grids and stores the analysis results in a new grid. Areas where the soil is sandy and the climate dry are marked with a "T," for True, because they pass Judy's test. Areas that fail the test are marked with an "F," for False.

Sometimes a true/false test is an oversimplification. Judy wants sandy soil, but she's willing to accept soil with some clay in it if the climate is very good. She needs a way to rank the sites from least to most desirable.

She does this by *reclassifying* the grids, or reassigning their cell values according to a common "goodness" scale. First she reclassifies the soil grid. Cells that are rich in clay are least desirable and get a "1." Clay soil with a little sand gets a "2," and sand with a little clay gets a "3." Sandy soil with no clay is best: it gets a "4." Next, she reclassifies the climate grid using the same scale.

The original soil and climate grids were overlaid and logically analyzed to see whether or not certain conditions were met. Now that both grids have been reclassified to measure suitability on a scale of 1 to 4, they can literally be added together. The area that has the highest score is the best location for growing oranges. The area with the second-best score is the second-best location, and so on.

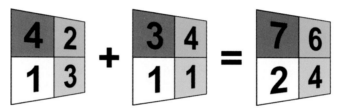

Judy adds the soil grid and the climate grid together to create an overall suitability grid. The higher the score, the more suitable the site.

This model is more complex than the original true/false test, but it still contains simplifying assumptions. It assumes, for instance, that soil and climate are equally important factors in the growth of oranges. It also assumes that changes in the quality of soil and climate are linear. In the real world, the difference between the best and second-best soil types might be larger or smaller than the difference between the second- and third-best types. Finally, as a matter of technique, the values in the overall suitability grid would normally be *weighted sums*, rather than simple sums, of the input grids. That way, the value scale can be kept constant throughout the model.

Its shortcomings notwithstanding, this example should give you a basic grasp of the modeling process. The next cases will introduce some refinements.

ENVIRONMENTAL SYSTEMS RESEARCH INSTITUTE, INC.

Creating a weighted distance model

A straight line may be the shortest distance as the crow flies, but it's not necessarily the most desirable route for someone on the ground. It may be a lot cheaper and easier, for instance, to build a road around a mountain rather than straight through it. Finding the best, or least-cost, path is a common problem in spatial analysis. By "weighting" distances—that is, incorporating factors such as the cost or difficulty of moving through terrain—it's possible to build a model of the best path between two points.

Steve, a Park Service employee, has been asked to design a bike path between points A and B. He has several issues to consider. The path can't go through the lake. It's undesirable to disturb the butterfly habitat. It's more time-consuming and expensive to build through forest than through grassland. To account for these factors, Steve will create a *weighted distance* model. He'll reclassify a terrain grid to assign difficulty values to each kind of terrain. Spatial Analyst will then be able to calculate the "cost" (measured in difficulty units) of traveling through each cell in the grid. This makes it possible to find the best path from A to B.

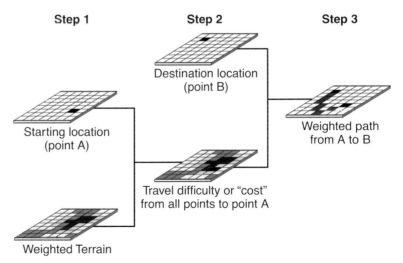

Step 1: Steve overlays a weighted terrain grid with a grid containing his starting point. Step 2: He produces a cost grid, in which each cell's value is the sum of the values of all other cells that are traveled through on the lowest-sum path to the starting point. He overlays the cost grid with a grid containing his destination point. Step 3: He produces a final grid showing the lowest-sum, or "least-cost," path from starting point to destination.

Steve's terrain grid shows the different kinds of land cover in the area. It also shows a lake and an environmentally sensitive butterfly habitat.

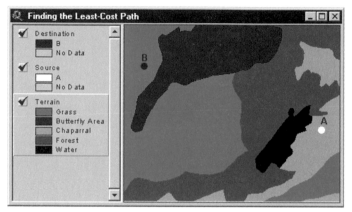

The best path from A to B must take into account barriers, such as the lake; impedances, such as dense vegetation; and environmentally sensitive land, such as the butterfly area.

Steve is ready to reclassify the terrain grid according to a scale that measures the "difficulty" of traveling, or building a path through, each cell. The difficulty scale can have any range that makes sense for the model. Steve chooses a scale of one to nine, where "1" represents the lowest difficulty and "9" the greatest.

It's easiest to build the path on grass, so grass is assigned a value of "1" in the reclassified grid. Chaparral is more difficult, because more labor is required to clear the land. Chaparral gets a "3." Forested land gets a "6." As for the butterfly habitat, Steve assigns it a value of "9," meaning that it's nine times as difficult to travel through a cell of butterfly area as through a cell of grassland. (The "difficulty" in this case represents a subjective value, concern for the environment, rather than an economic cost or a technical challenge.) The lake is reclassified as "No Data," which prohibits Spatial Analyst from putting a path through it. Steve's reclassified theme is called "Weighted Terrain" (step 1 of the model diagram).

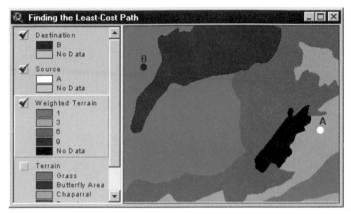

In the Weighted Terrain grid, the travel cost through one cell of grass equals the travel cost through three cells of chaparral, six cells of forest, or nine cells of butterfly area. (Spatial Analyst prefers a path through grass to a path through butterfly area until it becomes nine times as long.) The lake can't be traveled through.

Now that a difficulty value has been assigned to each cell in the Weighted Terrain grid, Spatial Analyst can calculate the total cost (cumulative difficulty) of reaching point A from any other cell in the grid (step 2 of the diagram). In the Cost grid, each cell's value is the cumulative travel cost of moving from that cell back to point A, along the path of lowest cost.

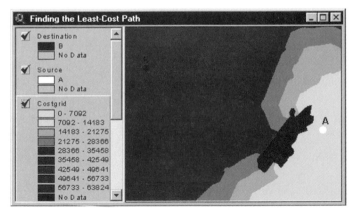

The cells in the Cost grid are classified by cumulative travel cost to point A. The higher the travel cost, the darker the shade of red. In a cell adjacent to point A, the cumulative travel cost would be "3." In a cell in the heart of the butterfly area, the cost would be around sixty thousand. Think of this as a long path of cell values whose sum is 9+9+9...+6+6...+3. Cells that can't be traversed are black.

Finally, Steve combines the Cost grid with a destination grid to find the least-cost bike path from A to B. The Result grid displays this path (step 3 of the diagram). As you might have expected, it drops down to the grassland and avoids heavily weighted cells, like the butterfly area.

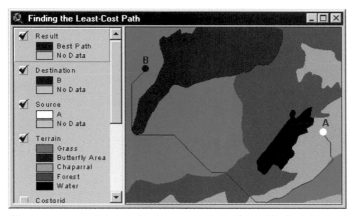

The least-cost path is not the shortest path, but it avoids the lake and butterfly areas, and minimizes the amount of forest that has to be crossed. The sum of the cell values traveled through on this path is the lowest of all possible paths.

All kinds of costs, values, and preferences can be built into a weighted distance model. In a full-scale model, the final weighted grid would be the product of several reclassified grids, each reflecting a different kind of cost or value.

Creating a site suitability model

In the bike path model, the cells in one grid were weighted according to difficulty. Now you'll look at a site suitability model, in which not only are the cells in each grid weighted, but the grids themselves are weighted relative to one another.

Eileen, a developer, is considering sites for a new amusement park. Below is a simplified diagram of the process she follows. She reclassifies the values in each of her grid themes on a suitability scale of one to nine (nine, in this case, being most suitable). Then she weights and combines the grids to create an overall suitability map.

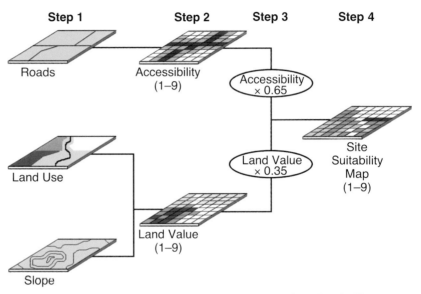

Step 1: Eileen begins with grid themes containing her raw data. Step 2: She reclassifies the Roads, Land Use, and Slope grids on a suitability scale of one to nine. She overlays Land Use and Slope to make a Land Value grid. Step 3: She applies a weight to each reclassified grid based on its importance to the model. Step 4: She produces an overall suitability grid theme.

Eileen's first step is to determine the criteria she'll use to evaluate sites. In this example, only accessibility from freeways and land value will be considered. (A full-scale model would include additional criteria.)

After selecting criteria, she gathers her raw data (step 1 of the model diagram). To analyze accessibility, she needs a theme of local freeways. To analyze cost, she needs two themes. A land use theme will give her information about the purchase cost of land (she knows that vacant land is cheap and developed land is expensive). A slope theme will tell her what kind of construction costs will be associated with grading the land.

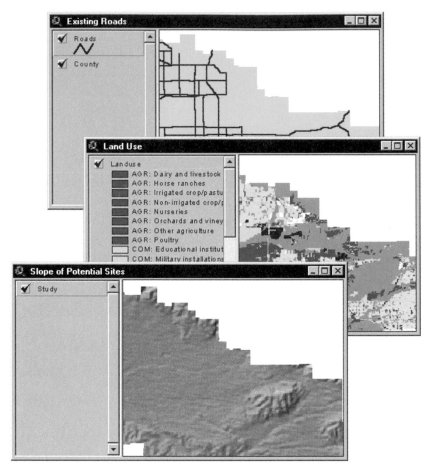

Eileen collects raw data showing roads, land use, and slope. The raw data can be in the form of feature, grid, or image themes.

After collecting her raw data, Eileen converts all the themes to grid themes, reclassifies them according to a common suitability scale, and combines them as needed (step 2 of the diagram). The values in the original grids may be in dollar amounts, distances, degrees of slope, and other units. Eileen translates them all onto a common scale of one to nine, with one being least suitable and nine most suitable for her purposes. (It's up to the modeler to choose the scale: it may have any range, and may be ascending or descending, as long as it's appropriate to the model.) As Eileen reclassifies each grid, she ranks the cells according to suitability (the closer to the freeway, the less developed, and the flatter, the better).

Using the roads data, Eileen creates a grid theme showing the average drive time from each cell to the nearest freeway off-ramp. Areas more than twenty minutes from an off-ramp are unacceptable. These areas get values of "No Data" so that Spatial Analyst won't even consider them. Areas less than twenty minutes away are ranked on a scale of one to nine according to their drive time (areas in the eighteen- to twenty-minute range get a one, areas in the zero- to two-minute range get a nine).

Eileen applies similar techniques to the land use and slope data, reclassifying both on her one-to-nine scale, and assigning "No Data" to cells with unacceptable values (residential land, for instance, or slopes that are too steep to grade).

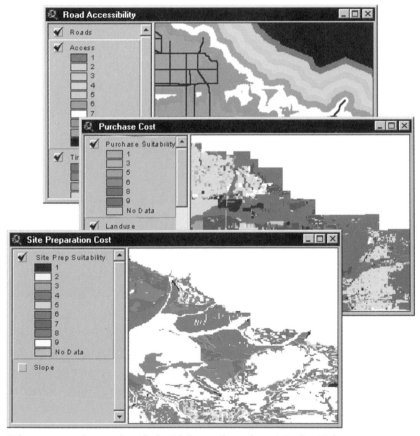

Eileen creates three reclassified grid themes from the original grid themes. Unacceptable areas are reclassified as "No Data." All other areas are ranked from one to nine in ascending order of suitability.

Now that the grids share a common suitability scale, Eileen can overlay them and apply some arithmetic. Still at step 2 of the diagram, she combines the Purchase Suitability grid (the reclassified land use grid) with the Site Preparation Suitability grid (the reclassified slope grid) to create an overall "land value" suitability grid.

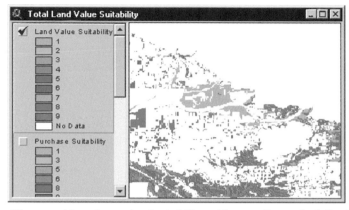

Eileen combines the two land suitability grids according to the following formula: (Purchase Suitability + Site Preparation Suitability)/2 = Land Value Suitability. In other words, corresponding cell values in the two grids are added together and averaged.

Next, Eileen has to decide how to weight the accessibility and land value grids relative to one another. To attract customers, it's important to site an amusement park near a freeway, but easily accessible land tends to be developed and expensive. Eileen is willing to pay more for land as long as it's close to a freeway, so, for her model, accessibility is a more important condition than cost. (How much more important? Two times? Five times? Eileen makes the decision, and Spatial Analyst applies it.) She ultimately decides that accessibility will make up 65 percent and land value 35 percent of the total suitability ranking (step 3 of the diagram). Spatial Analyst weights and adds the two grids to create a new overall suitability theme, called Potential Sites.

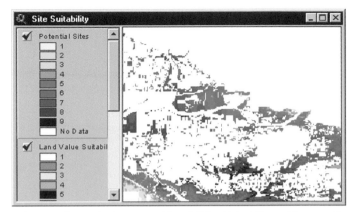

Eileen combines the Access and Land Value Suitability grids according to the following formula: (Access × 0.65) + (Land Value Suitability × 0.35) = Potential Sites. Highly suitable sites are blue to violet. Sites with low scores are yellow to green. White areas (No Data) are disqualified.

Eileen has reached step 4 of the diagram: she's created an overall site suitability map. She can continue to refine the model by making histograms and tables showing the amount of land available for each ranking. She can use Spatial Analyst to select an area large enough for the planned amusement park.

Expanding your GIS horizons

By now, you understand the power of grid-based themes. You've seen how Spatial Analyst creates surface maps from sample points, then uses those maps to calculate distances, slopes, probable paths, and much more. You've seen how, by counting cells and examining their values, you can make statistical tables, charts, and grids. And you've seen the immense value of Spatial Analyst as a decision support tool. By weighting and overlaying grids, you can build sophisticated models to choose optimal routes and sites and to simulate real-world processes.

You know what GIS is all about. You know how to use ArcView GIS and how to extend its capabilities with ArcView Network Analyst and ArcView Spatial Analyst. The world has come to your desktop.

What's next?

Environmental Systems Research Institute, Inc. (ESRI), builds quality GIS software products to meet a wide variety of application needs. Our product family covers the full range of GIS solutions, from general-use desktop GIS software to high-end systems used primarily for engineering and scientific applications.

ArcView® GIS software is the easy-to-learn, easy-to-use desktop GIS for everyone. ArcView's affordable price and its compatibility with other ESRI products make it ideal for those who want to get started with GIS. And ArcView runs on most computers, including PC, Macintosh®, and UNIX® and OpenVMS™ work-station platforms. (The multimedia applications and demonstration copy of ArcView provided with this book are intended for use on a PC. However, if you have a copy of ArcView on another platform, you can still do the exercises by following the instructions for Macintosh users or UNIX users in appendix D.)

For more information, give us a call at 1-800-GIS-XPRT (1-800-447-9778) (outside the United States, call 909-793-2853, extension 1235) or send E-mail to *info@esri.com*. Or, learn more about ArcView and other ESRI products and services (including ArcView classes) by connecting to our World Wide Web home page on the Internet: *www.esri.com*.

We look forward to hearing from you!

Corporate headquarters

ESRI
380 New York Street
Redlands, CA 92373-8100 USA
Telephone: 909-793-2853
Fax: 909-793-5953

Regional offices

Alaska
Telephone: 907-344-6613
Fax: 907-344-6813

Boston
Telephone: 508-777-4543
Fax: 508-777-8476

California
Telephone: 909-793-2853, extension 1906
Fax: 909-307-3025

Charlotte
Telephone: 704-541-9810
Fax: 704-541-7620

Denver
Telephone: 303-449-7779
Fax: 303-449-8830

ENVIRONMENTAL SYSTEMS RESEARCH INSTITUTE, INC.

Minneapolis
Telephone: 612-454-0600
Fax: 612-454-0705

Olympia
Telephone: 360-754-4727
Fax: 360-943-6910

St. Louis
Telephone: 314-949-6620
Fax: 314-949-6735

San Antonio
Telephone: 210-340-5762
Fax: 210-340-1330

Washington, D.C.
Telephone: 703-506-9515
Fax: 703-506-9514

International offices

ESRI–Australia
Telephone: 61-9-242-1005
Fax: 61-9-242-4412

ESRI–Canada
Telephone: 416-441-6035
Fax: 416-441-6838

ESRI–Europe
Telephone: 31-10-217-0690
Fax: 31-10-217-0691

ESRI–France
Telephone: 33-1-46-23-6060
Fax: 33-1-450-70560

ESRI–Germany
Telephone: 49-8166-380
Fax: 49-8166-3838

ESRI–Italy
Telephone: 39-6-406-96-1
Fax: 39-6-4069-6800

ESRI–Poland
Telephone: 48-22-256-482
Fax: 48-22-255-705

ESRI–South Asia
Telephone: 65-735-8755
Fax: 65-735-5629

ESRI–Spain
Telephone: 34-1-559-4375
Fax: 34-1-559-7071

ESRI–Sweden
Telephone: 46-23-84094
Fax: 46-23-84485

ESRI–Thailand
Telephone: 66-2-678-0707
Fax: 66-2-678-0321-3

ESRI–United Kingdom
Telephone: 44-1-923-210-450
Fax: 44-1-923-210-739

ESRI also has more than sixty distributors in other countries around the world. For more information, contact ESRI at 909-793-2853, extension 1235.

Data providers

Finding GIS data isn't hard; you just need to know where to look. You can get lists of GIS data vendors from a variety of sources. This is a list of lists, including some providers of free or low-cost data.

ArcData Catalog

This catalog is a collection of digital information products developed cooperatively by ESRI and more than thirty data providers. Included are data sets for such applications as business siting, market analysis, health care service, urban and transportation planning, education, and agriculture.

These data sets come from both public and private sources and include spatial, attribute, and image data. ArcData℠ Publishing Program data sets are compatible with ArcView® GIS software.

To receive your own copy of the *ArcData Catalog,* call 1-800-GIS-XPRT (1-800-447-9778) or direct E-mail to *info@esri.com.*

For more information about ArcData data sets and data publishers, visit our World Wide Web home page on the Internet at *www.esri.com*.

Other data resources

The Global Directory of Financial Information Vendors

James Essinger and Joseph Rosen. Business One Irwin (Publishers), 1994. ISBN 1-55623-788-X. A guide to more than 200 vendors of databases for the financial market.

World Mapping Today

R. B. Perry and C. R. Perkins. Butterworth & Co. (Publishers) Ltd., 1987. ISBN 0-408-02850-5.

A guide to maps available from around the world. It provides an overview of world mapping in the late 1980s and lists worldwide mapping agencies, graphic indexes, and catalogs of map publishers by country.

Manual of Federal Geographic Data Products

Published by the United States Federal Geographic Data Committee (FGDC).

Describes over 150 federal geographic products distributed by twenty-one U.S. federal agencies. Topographic data includes maps, imagery, and digital geographic locational data. It also includes data description, data coverage, delivery format, and ordering information.

Contact: Federal Geographic Data Committee Secretariat, United States Geological Survey, 590 National Center, Reston, Virginia 22092. Telephone: (703) 648-5514. Internet address: gdc@usgs.gov.

Statistical Abstract of the United States 1994

Published by the United States Department of Commerce, Economics and Statistical Administration, Bureau of the Census.

The standard summary of statistics on the social, political, and economic organizations of the United States. It includes comparative international statistics, foreign commerce and aid, and a guide to twenty-nine foreign statistical abstracts.

Contact: Customer Service, Bureau of the Census, Washington, D.C. 20233-8300. Telephone: 301-457-4100. Fax: 301-763-4794.

United States Census Bureau

The United States Census Bureau offers a variety of geographic tools, maps, reports, and tapes containing information about the population of the United States.

The Bureau reports on population and housing, economics, agriculture, and government censuses. A geographic database is also available: TIGER/Line Files (Topographically Integrated Geographic Encoding and Referencing).

Contact: Customer Service, Bureau of the Census, Washington, D.C. 20233-8300. Telephone: 301-457-4100. Fax: 301-763-4794.

GIS periodicals

Business Geographics

Contact: Business Geographics c/o GIS World, Inc., 155 East Boardwalk Drive, Suite 250, Fort Collins, Colorado 80525. Telephone: 970-223-4848. Fax: 970-223-5700. E-mail: BG@gisworld.com.

Geo Info Systems

Contact: Geo Info Systems, P. O. Box 6139, Duluth, Minnesota 55806-6139. Telephone: 800-346-0085, extension 226. Outside the United States: 218-723-9477.

GIS Europe

Contact: GeoInformation International, 307 Cambridge Science Park, Milton Road, Cambridge, CB4 4ZD, United Kingdom. Telephone: 44 181 402 8181. Fax: 44 181 402 8383.

GIS World

Contact: GIS World, Inc., 155 East Boardwalk Drive, Suite 250, Fort Collins, Colorado 80525. Telephone: 970-223-4848. Fax: 970-223-5700. E-mail: info@gisworld.com.

License Agreement

Important: Read carefully before opening the sealed media package

ENVIRONMENTAL SYSTEMS RESEARCH INSTITUTE, INC. (ESRI), IS WILLING TO LICENSE THE ENCLOSED SOFTWARE, DATA, AND RELATED MATERIALS TO YOU ONLY UPON THE CONDITION THAT YOU ACCEPT ALL OF THE TERMS AND CONDITIONS CONTAINED IN THIS LICENSE AGREEMENT. PLEASE READ THE TERMS AND CONDITIONS CAREFULLY BEFORE OPENING THE SEALED MEDIA PACKAGE. BY OPENING THE SEALED MEDIA PACKAGE, YOU ARE INDICATING YOUR ACCEPTANCE OF THE ESRI® LICENSE AGREEMENT. IF YOU DO NOT AGREE TO THE TERMS AND CONDITIONS AS STATED, THEN ESRI IS UNWILLING TO LICENSE THE SOFTWARE, DATA, AND RELATED MATERIALS TO YOU. IN SUCH EVENT, YOU SHOULD RETURN THE MEDIA PACKAGE WITH THE SEAL UNBROKEN AND ALL OTHER COMPONENTS TO ESRI.

ESRI License Agreement

This is a license agreement, and not an agreement for sale, between you (Licensee) and Environmental Systems Research Institute, Inc. (ESRI). This ESRI License Agreement (Agreement) gives Licensee certain limited rights to use the software and related materials (Software, Data, and Related Materials). All rights not specifically granted in this Agreement are reserved to ESRI and its Licensors.

Reservation of Ownership and Grant of License:

ESRI and its Licensors retain exclusive rights, title, and ownership to the copy of the Software, Data, and Related Materials licensed under this Agreement and, hereby, grant to Licensee a personal, nonexclusive, non-transferable, royalty-free, worldwide license to use the Software, Data, and Related Materials based on the terms and conditions of this Agreement. Licensee agrees to use reasonable effort to protect the Software, Data, and Related Materials from unauthorized use, reproduction, distribution, or publication.

Proprietary Rights and Copyright:

Licensee acknowledges that the Software, Data, and Related Materials are proprietary and confidential property of ESRI and its Licensors and are protected by United States copyright laws and applicable international copyright treaties and/or conventions.

Permitted Uses:

Licensee may install the Software, Data, and Related Materials onto permanent storage device(s) for Licensee's own internal use.

Licensee may make only one (1) copy of the original Software, Data, and Related Materials for archival purposes during the term of this Agreement unless the right to make additional copies is granted to Licensee in writing by ESRI.

Licensee may internally use the Software, Data, and Related Materials provided by ESRI for the stated purpose of GIS education.

Uses Not Permitted:

Licensee shall not sell, rent, lease, sublicense, lend, assign, time-share, or transfer, in whole or in part, or provide unlicensed Third Parties access to the Software, Data, and Related Materials or portions of the Software, Data, and Related Materials, any updates, or Licensee's rights under this Agreement.

Licensee shall not remove or obscure any copyright or trademark notices of ESRI or its Licensors.

Term:

The Agreement shall automatically terminate without notice if Licensee fails to comply with any provision of this Agreement. Licensee shall then return to ESRI the Software, Data, and Related Materials. The parties hereby agree that all provisions that operate to protect the rights of ESRI and its Licensors shall remain in force should breach occur.

Limited Warranty:

THE SOFTWARE, DATA, AND RELATED MATERIALS CONTAINED HEREIN ARE PROVIDED "AS-IS," WITHOUT WARRANTY OF ANY KIND, EITHER EXPRESS OR IMPLIED, INCLUDING, BUT NOT LIMITED TO, THE IMPLIED WARRANTIES OF MERCHANTABILITY AND FITNESS FOR A PARTICULAR PURPOSE.

ESRI does not warrant that the Software, Data, and Related Materials will meet Licensee's needs or expectations, that the use of the Software, Data, and Related Materials will be uninterrupted, or that all nonconformities, defects, or errors can or will be corrected. ESRI is not inviting reliance on the Software, Data, and/or Related Materials for planning or analysis purposes, and Licensee should always check actual data.

Limitation of Liability:

ESRI shall not be liable for direct, indirect, special, incidental, or consequential damages related to Licensee's use of the Software, Data, and Related Materials, even if ESRI is advised of the possibility of such damage.

No Implied Waivers:

No failure or delay by ESRI or its Licensors in enforcing any right or remedy under this Agreement shall be construed as a waiver of any future or other exercise of such right or remedy by ESRI or its Licensors.

Order for Precedence:

Any conflict between the terms of this Agreement and any FAR, DFAR, purchase order, or other terms shall be resolved in favor of the terms expressed in this Agreement, subject to the government's minimum rights unless agreed otherwise.

Export Regulation:

Licensee acknowledges that this Agreement and the performance thereof are subject to compliance with any and all applicable United States laws, regulations, or orders relating to the export of data thereto. Licensee agrees to comply with all laws, regulations, and orders of the United States in regard to any export of such technical data.

Severability:

If any provision(s) of this Agreement shall be held to be invalid, illegal, or unenforceable by a court or other tribunal of competent jurisdiction, the validity, legality, and enforceability of the remaining provisions shall not in any way be affected or impaired thereby.

Governing Law:

This Agreement, entered into in the County of San Bernardino, shall be construed and enforced in accordance with and be governed by the laws of the United States of America and the State of California without reference to conflict of laws principles. The parties hereby consent to the personal jurisdiction of the courts of this county and waive their rights to change venue.

Entire Agreement:

The parties agree that this Agreement constitutes the sole and entire agreement of the parties as to the matter set forth herein and supersedes any previous agreements, understandings, and arrangements between the parties relating hereto.

How to use the CD–ROM

The *Getting to Know ArcView GIS* CD–ROM features the Desktop GIS Primer, ArcView GIS Showcase, and ArcView GIS Tutorial multimedia applications. It also contains a copy of ArcView® GIS software that can be used to perform the exercises in chapters 7–27 of the book.

Desktop GIS Primer

Select this choice from the main screen of the application to see illustrations of the concepts presented in chapters 1–5 of the book.

ArcView GIS Showcase

Select this choice from the main screen of the application to see how ArcView software implements the concepts presented in chapters 1–6 of the book.

ArcView GIS Tutorial

Select this choice from the main screen of the application to start ArcView and begin doing the exercises in chapters 7–27 of the book. You can also use the ArcView GIS tutorial without the book. It contains a help system with all of the exercises from the book and "helper" videos that show you how to perform the tasks.

Installing the application

To install the application,

- Start Windows, place the Getting to Know ArcView GIS disc in the plastic disc holder (if required), and insert it into the CD–ROM drive. If you're running Microsoft® Windows® 95 or Windows NT™ version 4.0 or higher, choose **Run** from the Start menu. If you're running Windows version 3.1 or Windows NT version 3.51, open the Program Manager window and choose **Run** from the File menu.

- In the Command Line box, type the letter of your CD–ROM drive, a colon, a backslash, and the word SETUP (for example: **e:\setup**).

- Follow the instructions that appear on the screen.

The setup program will automatically install the required pieces of the Getting to Know ArcView GIS application. In part of the install you'll be asked to install optional components. These include the data used in the tutorial, the multimedia presentations, and the AVI "helper" videos. We *strongly recommend* that you install the data. Certain exercises can't be completed from the CD; they must be on your hard disk. The other components can be left on the CD, as long as the CD is in your CD–ROM drive when you run the application.

ENVIRONMENTAL SYSTEMS RESEARCH INSTITUTE, INC.

Starting the application

If you're running Windows 95 or Windows NT 4.0 or higher, the installation will create a Getting to Know ArcView GIS menu in the Program menu. With the CD still in the CD–ROM drive, select "Getting to Know ArcView GIS" from this menu.

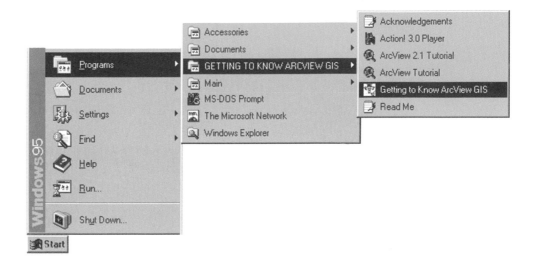

If you're running Windows 3.1 or Windows NT 3.51, the installation will create a Program Group and Windows icons for the application. With the disc still in the CD–ROM drive, start the application by double-clicking on the "Getting to Know ArcView GIS" icon.

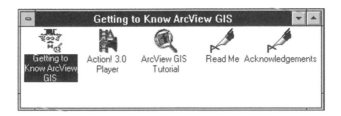

Using the application

After you've started Getting to Know ArcView GIS, you can select which of the three parts you'd like to see: the Desktop GIS Primer, the ArcView GIS Showcase, or the ArcView GIS Tutorial. You can also select Using this CD, which gives brief instructions on starting the primer, showcase, and tutorial, or Contacting ESRI, which lists ESRI's phone number and Web site information.

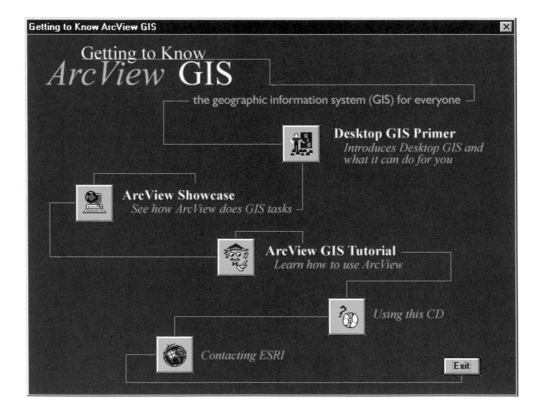

Using the Desktop GIS Primer

Click this button to start the Desktop GIS Primer. The primer contains six different presentations, which you can control with the buttons described below.

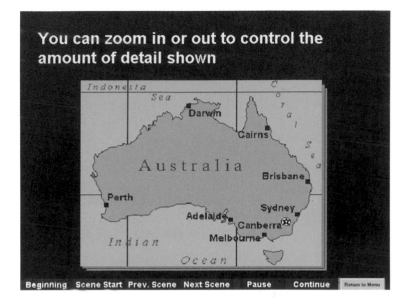

Beginning takes you to the beginning of the chosen section.

Scene Start takes you to the beginning of the current scene.

Prev. Scene takes you to the beginning of the previous scene.

Next Scene takes you to the beginning of the next scene.

Pause pauses the presentation.

Continue restarts the presentation after it's been paused.

Return to Menu stops the presentation and returns you to the Desktop GIS Primer menu.

Using the ArcView GIS Showcase

Click this button to start the ArcView GIS Showcase. The entire presentation will play unless you interrupt it by clicking one of the buttons on the left edge. You can interrupt the showcase at any time.

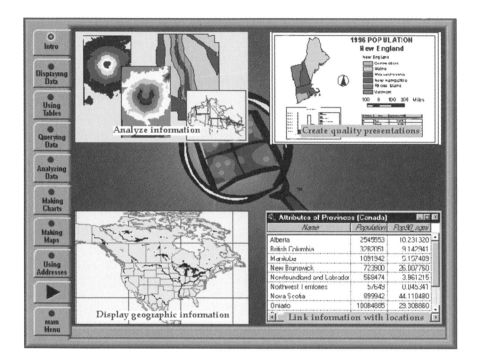

Click any one of the topic buttons along the left edge, such as Querying Data, or Making Charts, to play the presentation on that topic.

The Play/Pause button (second from the bottom) toggles between a right arrow (>) when paused and parallel lines (||) when playing.

The Main Menu button returns you to the main menu.

ENVIRONMENTAL SYSTEMS RESEARCH INSTITUTE, INC.

Using the ArcView GIS Tutorial

Click this button to start the ArcView GIS Tutorial. To use the tutorial, you must load the data for each exercise. If you're working from the book, step 1 of each exercise shows you how to load the data. Alternately, you can choose Tutorial Contents from the Help menu or click the button shown at the left to start the Getting to Know ArcView GIS help. You can then navigate to any chapter and exercise and work completely online.

In each exercise in the online help you'll see **Show Me** and **Load Data** buttons. Clicking **Show Me** starts a Windows video showing each step of the exercise in ArcView. Clicking **Load Data** loads the data for the current exercise into ArcView.

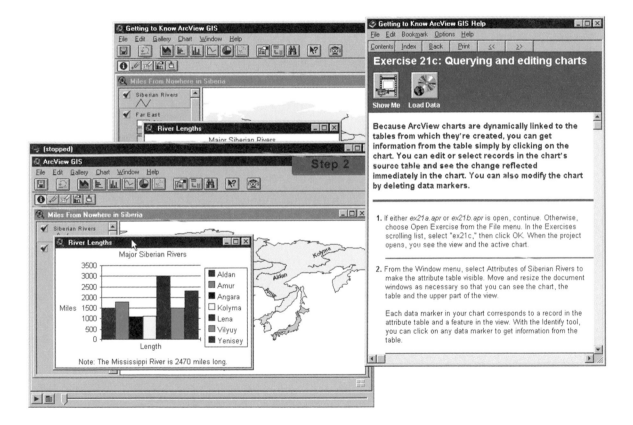

Improving the performance

This application has been designed to run well from any CD–ROM, but will run better from a triple- or quad-speed drive (or higher).

System requirements

- A multimedia PC, 486/33MHz

- 16MB of RAM

- 16MB of hard disk space

- CD–ROM drive

- 256-color display adapter

- Microsoft Windows version 3.1, Windows NT, or Windows 95 (see the sections below for minimal functionality on Macintosh or UNIX)

Macintosh users

The multimedia applications (Desktop GIS Primer, ArcView GIS Show-case, and ArcView GIS Tutorial) and the demonstration version of ArcView won't run on the Macintosh operating system. However, if you already have a standard copy of ArcView for Macintosh on your machine, you can still do the exercises in the book using the project data from the CD. First, you need to create a GTKAV folder, then copy the DATA folder from the GTKAV folder on the CD to the GTKAV folder on your volume. Next, you'll set two variables.

Open the ArcView:Etc:Startup file. At the beginning of the file, type these two lines, replacing *volume* with the name of the volume where you copied the GTKAV folder:

System.SetEnvVar ("GTKAVHOME", "*volume*:GTKAV:")

System.SetEnvVar("GTKAVDATA","*volume*:GTKAV:DATA:")

To do the exercises, start ArcView, then open the Gtkav.apr project located in the *volume*:GTKAV:DATA folder. Once this project is loaded, you can begin any exercise.

ENVIRONMENTAL SYSTEMS RESEARCH INSTITUTE, INC.

UNIX users

The multimedia applications (Desktop GIS Primer, ArcView GIS Showcase, and ArcView GIS Tutorial) and the demonstration version of ArcView won't run on a UNIX platform. However, if you already have a standard copy of ArcView on your machine, you can still do the exercises in the book using the project data from the CD. First, create a Gtkav directory on your drive (remember, UNIX is case-sensitive, so you must use an exact combination of capital and lowercase letters). Copy the Data subdirectory from the Gtkav directory on your CD to the Gtkav directory on your hard drive.

Add the following two lines to your .cshrc file, replacing *path* with the path to the Gtkav directory:

setenv GTKAVHOME *path*/Gtkav/

setenv GTKAVDATA *path*/Gtkav/Data/

Start ArcView and load the gtkav.apr project in the *path*/Gtkav/Data directory. Once you load this project, you can begin any exercise.

Technical support

While no direct technical support is available for the software on the CD, feel free to use either of our two self-help channels to get help with ArcView GIS: Fax-on-Demand and the World Wide Web. These services are free (other than provider charges and phone time), require no waiting, and are available twenty-four hours a day, seven days a week.

Fax-on-Demand

By simply dialing (909) 301-3111, users of ESRI desktop products can request that FAQ (Frequently Asked Questions) documents be faxed to a number anywhere in the United States and Canada. Full instructions are given for retrieving documents. Users can request an index of available documents or get up to three documents per phone call.

FAQs on the World Wide Web

Users with access to the Internet and the Web can now access FAQ documents on ArcView software. These are available in the Technical Support section of ESRI's World Wide Web home page and are updated at least once a week. Point your Web browser to *www.esri.com*.

Known problems

The CD–ROM has been thoroughly tested on many platforms and we have encountered one problem in the Desktop GIS Primer and the ArcView GIS Showcase. The combination of the following hardware, screen resolution, and number of colors causes the images in the presentation to appear as static:

Dell Dimension 466 DM (16MB RAM)
Diamond Viper Pro (2MB)
Dell S3 monitor
Windows 3.1
800 x 600 x 16.7M colors @ 75Hz

To correct this, reduce the number of colors to 256.

There may be problems with other hardware configurations. If you encounter any problems with the presentations themselves, we recommend that you change your screen resolution to 640 x 480 and the number of colors to 256. This is the target use configuration and should solve any problems. For the latest list of known problems, please read the **Read Me** file installed with the application.

ENVIRONMENTAL SYSTEMS RESEARCH INSTITUTE, INC.

Glossary

You'll find more information on many of these entries in the ArcView® GIS online help. You can also access an online help glossary by clicking the Find tab and typing **Glossary.**

address geocoding   The process of finding the map coordinates of a location from an address.

address matching   The process of finding correspondences between addresses in a table and the address attributes of a theme. Address matching is an essential part of address geocoding.

address range   The range of street numbers that occurs along a street or street segment. Typically, address ranges are stored as fields in the attribute table of a street theme and are used for geocoding.

alias   In ArcView, an additional name for a field in a table. Typically, an alias is a descriptive name for a field labeled with a code or an abbreviation.

application   A specific organization and use of GIS software tools to complete a task or manage a process. Broadly, a GIS project. Examples of GIS applications include finding efficient delivery routes, mapping wildlife habitats, and choosing optimal locations for businesses and facilities.

ARC/INFO   A GIS software package from ESRI. Coverages and grids created in ARC/INFO may be used in ArcView.

ArcView Network Analyst   An extension that gives ArcView the ability to analyze networks. Network analysis includes finding best routes, closest facilities, and facility service areas.

ArcView Spatial Analyst   An extension that gives ArcView the ability to create, query, analyze, and map cell-based raster data, and to perform integrated vector–raster analysis with feature-based and grid-based themes.

area feature   *See* **polygon feature.**

attribute   A piece of information describing a map feature. Attributes of a river might include its name, length, and average depth.

Avenue   The object-oriented programming language that comes with ArcView. Avenue provides tools for customizing ArcView and developing applications. *See* **object-oriented programming.**

buffer map   A map showing zones of a specified distance drawn around features.

CADReader extension   An extension to ArcView that allows CAD drawings to be added as themes.

CAD theme   A theme created from a CAD (Computer-Aided Design) file.

cell   The basic unit of spatial information in a grid theme. Cells are always square. A group of cells forms a grid. Each cell in a grid theme has an attribute value.

census block   The smallest entity for which the U.S. Census Bureau collects and tabulates decennial census information. A census block is bounded on all sides by visible features (for example, roads and streams) or nonvisible features (for example, township lines).

census tract   A statistical subdivision of a metropolitan area with between 2,500 and 8,000 inhabitants. Census tracts are designed by local committees to be relatively homogeneous with respect to population characteristics, economic status, and living conditions. They always nest within county boundaries and may be split by any subcounty geographic entity.

ENVIRONMENTAL SYSTEMS RESEARCH INSTITUTE, INC.

chart   A graphic representation of tabular data.

classification method   In ArcView, a formula for sorting attribute values into groups so that unique symbology can be assigned to each group. ArcView supports five classification methods: Natural Breaks, Equal Area, Equal Interval, Quantile, and Standard Deviation.

continuous surface map   A map representing a geographic phenomenon that lacks definite boundaries and has variable values across a surface (for example, elevation or temperature).

contour map   A map that represents elevation by a series of lines, with each line connecting a set of points of equal value. A contour map is a special case of an *isoline* map, where the phenomenon represented may be any measurable quantity, such as temperature or concentration of a solute. In ArcView, contour is used as a synonym for isoline.

coverage   An ARC/INFO data file in which geographic features are stored as points, lines, and polygons, and feature attributes are stored in associated INFO tables.

database   A collection of interrelated information, managed and stored as a unit, usually on some form of mass-storage system, such as magnetic tape or disk. A GIS database includes data about the spatial location, shape, and attributes of geographic features.

Database Themes extension   An extension that gives ArcView the ability to retrieve geographic information from a server machine running Spatial Database Engine™ (SDE™) software and add it as a theme.

data dictionary   A catalog describing the data stored in a GIS database. A data dictionary includes such information as the full names of attributes, meanings of codes, scale of source data, accuracy of locations, and map projections used.

data provider   A vendor who provides spatial data for use with GIS software.

dBASE file   A file format native to dBASE database management software. ArcView can read, create, and export tables in dBASE format.

decimal degrees Degrees of latitude and longitude expressed as a decimal rather than in degrees, minutes, and seconds. Decimal degrees are computed using this formula: Decimal Degrees = Degrees + Minutes/60 + Seconds/3,600. (73° 59' 15" longitude is equal to 73.9875 decimal degrees.)

demographics The statistical characteristics (for example, age, birth rate, and income) of human populations.

desktop GIS A geographic information system that runs on a desktop computer. *See* **geographic information system.**

destination table In ArcView, a table to which another table is joined or linked. The destination table is the table that is active at the time of the join or link. *Compare* **source table.**

digitizer A device consisting of a tablet and a cursor with crosshairs and keys. A digitizer converts the positions of features on a graphic image (like a paper map) to a series of x,y coordinates stored in computer files.

distance units The units (e.g., feet, miles, meters, or kilometers) ArcView uses to report measurements, dimensions of shapes, and distance tolerances and offsets. Distance units may be set independently of map units. *Compare* **map units.**

document A component of an ArcView project. Each document type (view, table, chart, layout, script) has its own window and interface.

dot density map A map in which dots are used to represent the density of an attribute (for instance, population).

equator A great circle on the earth equidistant at all points from the north and south poles. Its latitude is 0°.

ERDAS IMAGINE A format for storing images published by ERDAS.

event In ArcView, a location stored in tabular format. "34.03, 117.11" is an event (representing the latitude–longitude value 34.03° N, 117.11° W); so is "380 New York Street." ArcView can convert events into points (or line segments) on a map.

event theme   A theme created from a table of event locations.

extension   A program loaded inside ArcView to add new capabilities. Some extensions are provided with ArcView and others can be purchased from ESRI or third-party vendors.

extent   In ArcView, the geographic area displayed in a view window.

feature   A map representation of a geographic object. In ArcView, there are three types of features: points, lines, and polygons.

feature data   In a GIS, data in vector format representing geographic objects as points, lines, or polygons. *See* **vector format.**

field   A column in a table, containing the values for a single attribute. In ArcView, fields can be in numeric, string, date, or Boolean formats.

filtering   In ArcView, an operation that hides (without deleting) specified theme features in a view.

geocoding   *See* **address geocoding.**

geographic coordinates   Values of latitude and longitude that define the position of a point on the earth's surface. *See also* **spherical coordinate system.**

geographic data   Information about objects found on the earth's surface, including their locations, shapes, and attributes. Geographic data can be in vector, raster, or tabular format.

geographic information system   A configuration of computer hardware and software that captures, stores, analyzes, and displays geographic information.

GIS   Acronym for geographic information system.

global positioning system   A system that pinpoints locations on the earth's surface by using a receiving device to measure and triangulate distances from satellites.

GPS   Acronym for global positioning system.

graphic   In ArcView, an object drawn in a view that is not a feature and is not stored in a theme.

great circle　The line of intersection of the surface of a sphere and any plane passing through the center of the sphere. On the earth, all meridians and the equator are great circles.

grid theme　In ArcView, a theme in which geographic data is stored in an array of equally sized square cells arranged in rows and columns. Each cell has an attribute value. Grid themes can be created (or converted from feature or image themes) with the ArcView Spatial Analyst extension. *See also* **ArcView Spatial Analyst.**

histogram　A diagram showing the frequency (or count) of a given attribute.

hot link　In ArcView, a theme property that allows you, by clicking on a feature in a view, to display images or text files, open documents or projects, and run scripts.

image data　In a GIS, data in raster format, typically produced by an optical or electronic device. Satellite data, scanned data, and photographs are common forms of image data. *See* **raster format.**

INFO table　An attribute table associated with an ARC/INFO coverage.

interpolation　The process of determining unknown values that fall between known values. In a grid theme, the calculation of a cell's value based on the values of nearby cells.

JFIF　Acronym for JPEG File Interchange Format. A file format designed to contain JPEG-compressed images. A compressed image stored in a JFIF file can be exchanged between otherwise incompatible systems, such as Windows and UNIX. *See* **JPEG.**

join　In ArcView, an operation that appends the fields of one table to another table (usually a theme table) using a common field. *Compare* **link.**

JPEG　A standardized image compression mechanism designed to compress either full-color or gray-scale images of natural, real-world scenes. "JPEG" is an acronym for Joint Photographic Experts Group, the original name of the committee that wrote the standard. *See also* **JFIF.**

landmark theme   A point theme containing the locations of landmarks. The Network Analyst extension uses landmark themes to enhance travel directions for a route. *See also* **ArcView Network Analyst.**

latitude   A measurement, along a meridian, of the angle formed by straight lines drawn from the center of the earth to a point on the equator and to any point north or south. Latitude is 0° at the equator, 90° at the north pole, and –90° at the south pole. *Compare* **longitude.** *See* **equator; meridian.**

legend   A list of the symbols appearing on a map; a legend contains a sample of each symbol as well as text that interprets the symbol.

line feature   A line on a map representing a real-world object too narrow to be depicted as an area. Examples of line features include roads, rivers, and elevation contours.

link   In ArcView, an operation that relates two tables using a common field. When a record is selected in the destination table, all records with the same value in the common field are selected in the source table. Unlike a join, a link doesn't append the fields of one table to another. *Compare* **join.**

longitude   A measurement, along a parallel, of the angle formed by straight lines drawn from the center of the earth to a point on the prime meridian and to any point east or west. Longitude is 0° at the prime meridian, and is measured to 180° going east and –180° going west. *Compare* **latitude.** *See* **parallel; prime meridian.**

many-to-one relationship   In ArcView, a relationship between tables in which each record in a destination table corresponds to no more than one record in a source table, and in which correspondences need not be unique (many records in the destination table may correspond to a single record in the source table). An example of a many-to-one relationship is that between a table of land parcels and a table of owner names. Tables in a many-to-one relationship may be joined or linked. *Compare* **one-to-many relationship; one-to-one relationship.**

map   A graphical representation on a planar surface of the physical features of a portion of the earth's surface.

map projection A mathematical formula that converts spherical coordinates of latitude and longitude to planar coordinates on a map. Map projections distort one or more of these spatial properties: distance, area, shape, direction.

map units The coordinate units (for example, decimal degrees, meters, or miles) in which spatial data is stored.

meridian A great circle passing through the north and south poles of the earth. *See* **great circle.**

neatline A border drawn around a map.

network In ArcView, an interconnected set of lines representing possible paths from one location to another. A city streets theme is an example of a network.

normalization In ArcView, the division of one set of numeric attribute values by another to obtain a ratio. (For example, to find out population density, you would normalize population by area.) Also, the division of each attribute value in a set by the sum of the values in that set to obtain a percentage of total.

north arrow A map component that shows a map's orientation.

object-oriented programming A computer programming model characterized by the use of objects and messages (messages are called *requests* in Avenue). An *object* is a data construct having specific properties with changeable values. A *message* is an instruction to an object to change the value of one of its properties. In ArcView, for example, you might draw a rectangular graphic in a view. This graphic is an object with a unique set of properties: it has boundaries, a location in the view, a color, a fill pattern, and other characteristics. Avenue requests (messages) can be sent to this object to select or unselect it, to change its color, and so on. *See also* **Avenue.**

one-to-many relationship In ArcView, a relationship between tables in which a given record in a destination table may correspond to many records in a source table. An example of a one-to-many relationship is that between a table of office buildings and a table of building occupants. Tables in a one-to-many relationship may be linked, but should not usually be joined. *Compare* **many-to-one relationship, one-to-one relationship.**

one-to-one relationship	In ArcView, a relationship between tables in which a given record in a destination table corresponds to no more than one record in a source table, and in which each correspondence is unique (no two destination table records correspond to the same source table record). An example of a one-to-one relationship is that between a table of states and a table of state capitals. Tables in a one-to-one relationship may be joined or linked. *Compare* **many-to-one relationship, one-to-many relationship.**
operator	A mathematical or logical function used in queries (for example, "greater than," "not").
origin	A point in a coordinate system that serves as a reference for defining other positions in the system. In a planar coordinate system, the origin (commonly, 0,0) is the point at which the x- and y-axes intersect; in the earth's spherical coordinate system, the origin ($0°, 0°$) is the intersection of the equator and the prime meridian.
parallel	A circle on the earth that is parallel to the equator and connects points of equal latitude.
parcel	An area of land whose boundaries have been surveyed and recorded.
planar coordinate system	A two-dimensional measurement system that defines locations on a map based on their distance from an origin (0,0) along two axes, a horizontal x-axis representing east–west and a vertical y-axis representing north–south. *Compare* **spherical coordinate system.**
planimetric map	A map that presents the horizontal, but not the vertical, positions of the features represented. *Compare* **topographic map.**
point feature	A point on a map representing a geographic object too small to show as a line or polygon. Examples of typical point features include wells and fire hydrants. (Since map scale affects feature representation, the same city might be a polygon feature on a large-scale map and a point feature on a small-scale map.)
polygon feature	A polygon on a map representing a geographic object too large to be depicted as a point or line. Examples of polygon features include census tracts, lakes, and countries.

prime meridian A great circle passing through the north and south poles and through Greenwich, England. Its longitude is 0°.

project In ArcView, a file for organizing related documents. A project typically brings together all the views, tables, charts, layouts, and scripts that are used to complete a task or manage a process.

projection *See* **map projection.**

project repair In ArcView, the process of updating document pathnames when referenced data is moved from one disk location to another, or when the name of a data source or pathname component changes.

query In ArcView, a logical statement used to select features or records. A simple query contains a field name, an operator, and a value.

raster format In a GIS, a cell-based representation of map features. Each cell in the structure has a value; a group of cells with the same value represents a feature. Images and grids are stored in raster format. *Compare* **vector format.**

record A row in an ArcView table. If the table is a theme table, each record corresponds to a map feature.

route In a GIS, a path through a network. *See* **network.** *See also* **route cost.**

route cost The measurement of a route in terms of distance, time, or another parameter. The route cost is used to determine the most efficient route through a network. *See* **network; route.**

scale The relationship between the dimensions of features on a map and the geographic objects they represent on the earth, commonly expressed as a a ratio or fraction. A map scale of 1:100,000 means that one unit of measure on the map equals 100,000 of the same unit on the earth.

scale bar A map component that graphically shows a map's scale.

scanned data Information, such as a photograph or document, that has been converted from printed to digital format.

script In ArcView, a program written in the Avenue scripting language. *See* **Avenue.**

shape In general, the visible form of a geographic object. In ArcView, the technical name for a map feature stored in shapefile format. *See* **shapefile.**

shapefile ArcView's format for storing the location, shape, and attribute information of geographic features.

source table In ArcView, a table that is joined or linked to another table. The source table is the table that is inactive at the time of the join or link. *Compare* **destination table.**

spatial analysis In ArcView, the determination of spatial relationships among map features through the use of spatial join and theme-on-theme selection. Several different spatial relationships can be analyzed: *containment* (features completely or partially contain, or are contained by, other features); *intersection* (features intersect other features); *adjacency* (features touch other features); and *proximity* (features lie within a specified distance of other features). *See* **spatial join; theme-on-theme selection.**

spatial data Information about the locations and shapes of objects found on the earth's surface. Spatial data is a subset of geographic data (the latter also includes attribute information). Spatial data can be in vector, raster, or tabular format. *See* **attribute; geographic data.**

Spatial Database Engine (SDE) A database product from ESRI that employs client/server architecture and a set of software services to manage large, shared geographic data sets and to perform spatial operations on them. ArcView, running on a client machine, can request data from the database server. *See also* **Database Themes extension.**

spatial join In ArcView, an operation that joins two theme tables on the Shape field. When a polygon theme is spatially joined to a point theme, the attributes of each polygon in the source table are appended to all points in the destination table that are contained by that polygon. When a point theme is spatially joined to another point theme, the attributes of each point in the source table are appended to the nearest point in the destination table. *See also* **spatial analysis.**

spherical coordinate system   A two-dimensional system that defines locations on the surface of a sphere based on angular measurements from the center of the sphere to a primary great circle (for example, the equator or the prime meridian) and a secondary great circle (a given meridian or the equator) whose plane is perpendicular to the first. *Compare* **planar coordinate system.**

summarize   In ArcView, an operation that counts the occurrences of each unique value in a highlighted field of a table. The results are saved to a new table, called a *summary table*. The summary table contains one record for each unique value counted and fields with the names of the values and the number of occurrences of each. The summary table may also contain statistical information.

summary table   *See under* **summarize.**

symbol   A particular graphic element (defined by some combination of shape, size, color, angle, outline, and fill pattern) used to draw a feature. For example, a point theme of hospitals might be symbolized with red crosses.

table   In ArcView, a data structure that stores attributes in rows and columns. Also called an *attribute table*. ArcView can read tables in dBASE, INFO, or text file format. *See also* **theme table.**

tabular data   In a GIS, locational or attribute information about geographic objects stored in a table.

text label   Text added to a map to help identify a feature.

theme   In ArcView, a set of geographic features of the same type, along with their attributes. Examples of themes are Atlanta points of interest, Redlands land parcels, and world rivers. A theme is stored as a unique set of files.

theme-on-theme selection   In ArcView, an operation that selects map features on the basis of a specified spatial relationship to other map features (which may or may not be in the same theme). *See also* **spatial analysis.**

theme table   A table associated with a particular theme, in which each record corresponds to one theme feature. A theme table has a Shape field that displays the feature type (for instance, polygon) of each feature in the theme and also contains its location information. *See also* **table.**

TIGER   Acronym for Topologically Integrated Geographic Encoding and Referencing. A data format used by the U.S. Census Bureau to store street address ranges and census tract/block boundaries.

topographic map   A map that displays both the horizontal and vertical positions of the features represented; distinguished from a planimetric map by the addition of relief in measurable form. A topographic map uses contours or other symbols to represent mountains, valleys, and plains. *Compare* **planimetric map.**

trade area   A geographic zone containing the people who are likely to purchase a firm's goods or services.

vector format   In a GIS, a coordinate-based representation of map features. A point is stored as a single x,y coordinate, a line as a pair of x,y coordinates, and a polygon as a set of x,y coordinates, each of which marks a vertex of the polygon. *Compare* **raster format.**

vertex   The point at which two sides of a polygon meet.

x-axis   In a planar coordinate system, the horizontal axis representing east–west.

x,y coordinates   Values representing the location of a point in a plane, relative to the axes of its coordinate system.

y-axis   In a planar coordinate system, the vertical axis representing north–south.

Sources consulted in preparing this glossary

Print

Definitions of Surveying and Associated Terms. Joint Committee of the American Congress on Surveying and Mapping and the American Society of Civil Engineers. 1984.

The Dictionary of Marketing. ed. Rona Ostrow. Fairchild Publications, New York, New York. 1988.

Digitech Systems' Dictionary of Facility Management and Geographic Information Systems Terminology. Digitech Systems, Indianapolis, Indiana. 1991.

Map Projections. Environmental Systems Research Institute, Inc., Redlands, California. 1994.

Microsoft Press® Computer Dictionary. Microsoft Press, Redmond, Washington. 1991.

Programming with Avenue. Environmental Systems Research Institute, Inc., Redlands, California. 1995.

Understanding GIS—The ARC/INFO Method. Environmental Systems Research Institute, Inc., Redlands, California. 1995.

Webster's Ninth New Collegiate Dictionary. Merriam-Webster, Inc., Springfield, Massachusetts. 1991.

Electronic

Random House Personal Computer Dictionary—Electronic Version. Philip E. Margolis. 1996.

Online

The GIS Glossary from ESRI
http://www.esri.com

All About GPS
http://www.trimble.com

University of Michigan Documents Center, 1990 Census of Population and Housing
http://www.lib.umich.edu

U.S. Census Bureau
http://www.census.gov

Index

ENVIRONMENTAL SYSTEMS RESEARCH INSTITUTE, INC.

ArcView GIS GUI Quick Reference

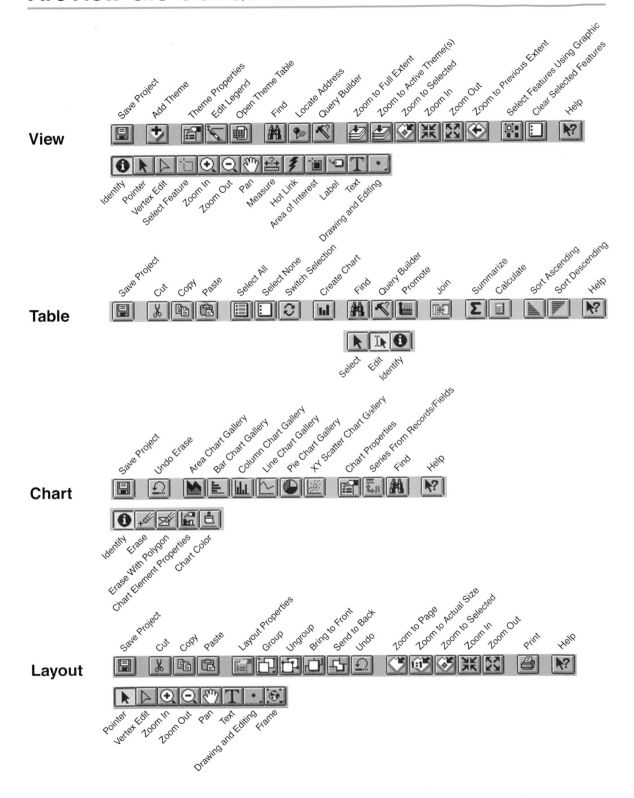

Getting to Know ArcView GIS workbook evaluation

After you complete the *Getting to Know ArcView GIS* workbook, we'd like to hear from you. You can help us improve the next edition by completing this questionnaire and returning it to us with your comments and suggestions.

GENERAL INFORMATION

1. What type of organization do you belong to?
   - a. Educational institution
   - b. Large business (50 employees or more)
   - c. Small business (fewer than 50 employees)
   - d. Government agency (federal/state/local)
   - e. Other _____
   - f. Not affiliated with an organization

2. What kind(s) of applications do you work with?
   - a. Urban/Municipal
   - b. Environmental/Natural resources/Forestry
   - c. Mapping and cartography
   - d. Transportation
   - e. Incident mapping for police, fire, health, or other services
   - f. Oil/Gas/Minerals
   - g. Facility management of roads and utilities
   - h. Other _____
   - i. No specific application area

3. What kind of computer and operating system did you use *Getting to Know ArcView GIS* with? (for example, 486/33 computer with 24MB RAM and Microsoft® Windows® 95)

   _____

ABOUT THE WORKBOOK

1. How do you rate part 1 (chapters 1–6) of *Getting to Know ArcView GIS*?
   ☐ Excellent ☐ Good ☐ Fair ☐ Poor

2. How do you rate part 2 (chapters 7–29) of *Getting to Know ArcView GIS*?
   ☐ Excellent ☐ Good ☐ Fair ☐ Poor

3. Did the text explain the concepts in a clear and logical way?
   ☐ Always ☐ Usually ☐ Sometimes ☐ Rarely
   If there were concepts you thought were poorly explained, what were they?

   _____

4. How easy or difficult did you find the exercises?
   ☐ Too easy ☐ About right ☐ Too difficult

5. Were there any exercises that were particularly difficult or otherwise unsatisfactory?
   ☐ Yes ☐ No
   If yes, which ones were they? _____

6. Did you read the entire book and complete all the exercises?
   ☐ Yes ☐ No If not, which chapters or exercises did you skip? _____

7. How much time did you spend *Getting to Know ArcView GIS*?
   About _____ hours

BUSINESS REPLY MAIL

FIRST-CLASS MAIL   PERMIT NO. 615   REDLANDS, CA

POSTAGE WILL BE PAID BY ADDRESSEE

Educational Products

ESRI

380 New York Street

Redlands, CA 92373-9870

ABOUT THE CD

1. Did you have any difficulties installing the CD?
   ☐ Yes   ☐ No   If yes, what were they?_____

2. Were there any exercises that didn't work properly?
   ☐ Yes   ☐ No   If yes, which ones were they? _____

3. Did you find the Desktop GIS Primer useful?
   ☐ Very   ☐ Somewhat   ☐ Not very

4. Did you find the ArcView GIS Showcase useful?
   ☐ Very   ☐ Somewhat   ☐ Not very

5. Did you find the ArcView GIS Tutorial useful?
   ☐ Very   ☐ Somewhat   ☐ Not very

Please add any comments that will help us improve the book: